Hop Production

Developments in Crop Science

Developments in Crop Science 16

Hop Production

Edited by
Václav Rybáček
Faculty of Agronomy, Agricultural University, Prague, Czechoslovakia

ELSEVIER

Amsterdam – Oxford – New York – Tokyo 1991

List of authors:

Prof. Ing. Václav Fric, Ph.D., Prof. Ing. Josef Havel, Ph.D., Ing. Vladimír Libich, Ph.D., Ing. Josef Kříž, Ph.D., Ing. Karel Makovec, Ph.D., Ing. Zdeněk Pertlík, Ph.D., Prof. Ing. Václav Rybáček, Ph.D., Ing. Jaromír Sachl, Ph.D., Ing. Antonín Srp, Ph.D., Ing. Josef Šnobl, Ph.D., Ing. Miroslav Vančura

Editor:

Prof. Ing. Václav Rybáček, Ph.D.

Scientific editors:

Prof. Ing. dr. Ladislav Hruška, Ph.D., prof. Mikuláš Klapal

Drawings:

Otakar Procházka

Revised translation of the Czech edition Chmelařství published by the State Agricultural Publishing House – Prague. Czechoslovakia

Distribution of this book is being handled by the following publishers:
For the United States and Canada
Elsevier Science Publishing Company, Inc.
655 Avenue of the Americas
New York, N.Y. 10010

for the East European Countries, China, Cuba, North Korea, Mongolia and Vietnam
State Agricultural Publishing House
Václavské nám. 47
Praha 1, Czechoslovakia

for all remaining areas
Elsevier Science Publishers
25 Sara Burgerhartstraat
P.O. Box 211, 1000 AE Amsterdam, The Netherlands
Library of Congress Cataloging in Publication Data

Chmelařství. English
Hop production/edited by V. Rybáček.
p. cm. – (Developments in crop science; 16)
Includes bibliographical references.
ISBN 0-444-98770-3
1. Hops. 2. Hops–Harvesting. 3. Hops–Economic aspects.
I. Rybáček, V. (Václav) II. Title. III. Series.
SB317.H64C4713 1990
633.8'2–dc20   89-26275 CIP

ISBN 0-444-98770-3 (Vol.16)
ISBN 0-444-41617-X (Series)

Joint edition published by
Elsevier Science Publishers, Amsterdam, The Netherlands and SZN, State Agricultural Publishing House, Prague, Czechoslovakia

Copyright © 1991 by Státní zemědělské nakladatelství, Praha
Translated by Jaromír A. Máša, Ph.D.,
Transferred to digital printing 2006
Printed and bound by CPI Antony Rowe, Eastbourne

CONTENTS

7

INTRODUCTION

The utilization of new knowledge in natural and social sciences is important not only for the scientific and technical progress of hop production but also for that of other crops. Crop production reflects the biological, technical and economic aspects of production technology, and it must also respect, and apply, new scientific knowledge. Conversely, when new urgent problems arise in production technology they should be immediately included in the programmes of scientific and technical research.

In hops, new knowledge results from scientific research, from scientifically verified field experiments, and from accurate observations in hop production. This knowledge represents the starting point of the common procedure: research – production – application. The short-term and long-term trends in other fields of research into plant production should be considered not only in relation to current and forecast trends in hop production but also to future hop research.

Hop cones constitute the main harvest product, and because their quality and their applicability to the brewing process are the main concern of hop production, the chapter on the application of hop products precedes the chapters on hop production.

In order to make particular aspects of production technology more easily understood, there are separate chapters entitled "Biological bases of production" and "Organization and economics of production". Technical operations, especially those involving mechanization and the application of chemicals are concerned directly in the technology of hop production, a term which also includes the various biological and economical components. These basic components of production technology must be considered in their complex interrelationships involving the whole production process and in their non-interchangeability, i.e. the state of their equivalency.

Clearly the fundamental basis of production technology resides in the biological properties of the hop plant. This does not mean, however, that the whole production technology is totally governed by the biological requirements of the hop plant. On the contrary, the technical and economic aspects of hop production require certain deviations from the optimal require-ments. The ability to make these deviations in favour of large scale production technology is limited by the potential adaptability of the hop. Where the tolerable adaptability is exceeded and the basic biological requirements of the plant are not satisfied, then the whole basis of production is threatened, because this depends almost entirely on the production capacity of hop vegetation.

Any non-adherence to the technical and economical aspects of production is clearly reflected in the efficiency of production. Thus a continuous inefficiency endangers productivi-ty in direct relation to the production capacity of the crop. The maximum harmony of those biological, technical and economic aspects characteristic of the whole production process will achieve the maximum yield from the planned hop production.

CHAPTER I
CULTIVATION AND UTILIZATION OF HOP PRODUCTS

The hop cone is, without doubt, the most important raw material for beer production due to the bitter taste which it imports. Moreover, since the bitterness of the beer is a dominant factor in its acceptability it follows that the hop is decisive in the final taste of beer. The hop cone has many components, but not all of them are involved in beer technology nor in the final product.

MAIN COMPONENTS OF HOP CONES

Lupulinic resins

From the standpoint of brewing, the most important substances in the cones of hop are lupulinic resins, as they are the cause of the bitter taste. These resins are a mixture of sparingly soluble substances. Shown below is a basic classification of lupulinic resins followed by a description of their features in an attempt to clarify a complicated situation.
Classification of lupulinic resins:
.1. soft resins: 1.1. α-bitter acids (humulones);
                             1.2. β-bitter acids (lupulones);
                             1.3. γ-bitter acids (humulinones);
                             1.4. δ-bitter acids (hulupones);
                             1.5. ε-bitter acids (not yet specified);
2. hard resins: 2.1. δ-resins;
                             2.2. γ-resins.

As shown in this classification, it is possible to divide the lupulinic resins into two main fractions, and from the standpoint of brewing the most valuable is the group of soft resins.

The soft resins

The soft resins include many substances which contribute to different extents to the final bitterness of the beer. The most important are the optically active α-bitter acids (humulones). They are a mixture of five different homologues characterized by different radicals in the second position of a benzene nucleus. Their oxidation produces γ-bitter acids (humulinones) with a pentacyclic structure and they similarly consist of five different homologues. They are less bitter than the α-bitter acids. The ultimate oxidation products of α-bitter acids are the δ-resins which are included in the hard resins. Their bitterness ranges from 15 to 20 per cent of that of the α-bitter acids.

The β-bitter acids are in fact not bitter. Chemically they again consist of a mixture of five homologues. Unlike the α-bitter acids they are not optically active.

12

The oxidation of β-acids produces δ-acids (hulupones) with a pentacyclic structure. Unlike their precursors, the β-bitter acids, these compounds are bitter and make an important contribution (30 to 50 per cent) to the final bitterness of the beer. Their further oxidation produces insoluble γ-resins lowering the value of hop for brewing. This resin fraction too is included in the group of hard resins.

There are different levels of bitterness among certain non specified soft resins. This fraction includes transformation products of the original bitter acids with as yet unknown structures.

The hard resins

The hard resins had for long been assumed to be inactive ballast. Only later was it shown that the members of one fraction of these resins, the so-called δ-resins, are very soluble and are bitter.

Czech hops, as distinguished from those of foreign provenance are typically well-endowed with α-bitter acids and the β-fraction (a complex of soft resins without the α-bitter acids) in the proportion of 1:1.2 to 1:1.5 i.e. more of the β-fraction than the α-bitter acids.

The α-bitter acids make their impact when fresh hops are used and their effect as the bitter substances in the beer decreases with increasing duration of storage. With regard to the quality of bitterness, our investigations have shown that unlike the β-fraction, the α--bitter acids have a tendency to produce rough and pronounced bitterness in the beer. We have verified, in numerous examinations, that the β-fraction is less efficient than the α-bitter acids in terms of quantity of bitterness, but it produces a gentle bitterness. This leads to the conclusion that hops with a higher content of β-fraction are better as a raw material for production of beers with a fine harmonious bitterness. The soluble hard resins, the so-called δ-resins, are also bitter. Their contribution to the total bitterness of the beer is, however, the least and depends on the content of the resin fraction in the hops. This content increases with the increasing length of the storage period. Qualitatively, the bitterness of the δ-resins is regarded as rough and crude.

This short survey has indicated the importance and effectiveness of individual resin fractions in imparting bitterness to beer. Quantitatively, the α-bitter acids are foremost, followed by the β-fraction and finally the δ-fraction, but qualitatively the β-fraction leads the α-bitter acids followed by the γ-resins. The α-bitter acids are very sparingly soluble and only in prolonged boiling are they transformed to very soluble and intensely bitter iso-α-acids. This transformation is effected by many factors, particularly by the intensity of boiling and by the pH. The original β-acids unlike the α-bitter acids are not transformed to their isomers by boiling. The β-bitter acids are transformed by polymerization and oxidation to products which are bitter and highly soluble in the fresh mash. These transformations proceed to a small extent during maturation of hop and increasingly during storage and the boiling process. Non-specified soft resins and δ-resins are soluble and are taken up by the fresh mash during the boiling stage.

The various components of lupulinic resin affect the *brewing qualities* of the hop as measured by its *contribution of bitterness*. Authorities differ in their evaluation of the effectiveness of the various components and consequently there are many formulae for the use of the hop in brewing. Mentioned below are some of the formulae used to determine the amount of hops to be used in brewing.

The first formula was derived by WÖLLMER (1925) as follows:

value of bitterness = $\alpha + (\beta/9)$,

where α = quantity of α-bitter acids in per cent;

β = quantity of the β-fraction in per cent.

In practice it was found, that this formula could be used only for fresh hops but is unsatisfactory for hops whose content of hard resins exceeds 15 per cent. Therefore, KOHLBACH (1939) proposed another formula to calculate the bitterness for older hops:

$$\text{value of bitterness} = \frac{a(100 - 0.4b)}{100 - 2.2b}$$

where a = value of bitterness according to Wöllmer;
$\quad b$ = content of hard resins in per cent minus 15.

This formula, however, did not find broad application. It is clear that both formulae, especially Wöllmer's assume, that the α-bitter acids account for the bitterness of beer, but worldwide research has shown, that this is not correct. SALAČ and DYR (1944) for example, proved, that the bitterness of β-fraction is not 1/9 of the α-bitter acids as given in Wöllmer's formula (1925), but is 1/3. Therefore, the formula was modified by these authors (1944):

$$\text{value of bitterness} = \alpha + \frac{\beta}{3}$$

Similar conclusions were reached by TOMBEUR and DE CLERCK (1934) and MIK-SCHIK (1963). New knowledge about the properties of the hard resins led to more suitable formulae, for the expressions of hop bitterness. According to VANČURA and BEDNÁŘ (1963) this value can be objectively expressed according to the quantity of so called "effective bitter substances", calculated as follows:
$\quad$ effective substances (%) = MP (%) + δ (%),
where MP = quantity of soft resins in per cent;
$\quad \delta$-resins are also in per cent.

For the purposes of this formula the quantity of γ-resins can be determined by calculation based on current values, according to Wöllmer's analysis (WÖLLMER, 1925). Dosing with hops according to this formula was confirmed in operative conditions as being correct.

One of the most recent methods for the establishment of the hop bitterness is that of PFENNINGER and SCHUR (1968). These authors give an expression for *general bitterness* as the milligrams of bitter substances present in a solution made from 1 gram of hop. Under laboratory conditions, a sample of hop is boiled for sixty minutes in buffer solution at pH 5.55. Bitter substances are extracted from this solution by chloroform and their quantity established by spectrophotometry. General bitterness is calculated according to the formula:

$$\text{general bitterness (mg.g}^{-1}) = \frac{E_{279} \times 115}{n}$$

where E^{279} = extinction of chloroform extract at wave length 279 nm;
$\quad n$ = mass of the sample in grams.

Unlike the previous method, this method involves better laboratory equipment (spectrophotometer).

These formulae for the value of bitterness of hop indicate its importance for the production of beer with standardized and reproducible bitterness. Because the bitterness of the same hop changes during the storage and because different hops are distinguished by different values, the objective analytical establishment of bitterness is increasingly important and its application to the brewing industry grows. According to the investigation of the Brewing Research Institute in Zürich (Switzerland) almost 50 per cent of the breweries questionéd in 1973 used hops at a rate calculated according to the analytically determined value of their bitterness.

Lupulinic tannin – polyphenol substances

Another component of hop, important in brewing, is the lupulinic tannin. This is a mixture of polyphenolic substances whose main components are anthocyanins and leucoanthocyanins, flavonols and catechins. This mixture is soluble in water and reacts very actively with malt proteins. Its *effect on the production of beer (Czech type)* is as follows:

1. It positively affects the strength and quality of the beer as well as the boiling process when it has a clearing effect on medium-molecular proteins. At the same time it holds the high molecular proteins in solution.

2. It promotes complex formation between proteins and bitter substances. There is thus a verified stabilizing effect of lupulinic tannin on bitter substances.

3. It participates to a large extent in the formation of the flavour regarded as typical of Czech beers.

There are, however, certain negative properties attributed to the lupulinic tannin. It has, for example, an unfavourable effect on the colloid stability of the beer. It should be mentioned here that STEINER (1965) suggested, that only 10–20 per cent of the total polyphenol substances in beer originates from the hop. The colloidal stability of the beer is probably more affected by other polyphenol substances derived mainly from malt.

Hop essential oil

Hop oil consists of a mixture of hydrocarbons and oxygenous complexes of terpene series. This mixture is mostly composed of the hydrocarbon fraction (60–70 per cent) containing such important essential oils as humulene, myrcene, farnesene and caryophyllene. The most important oils in the oxygenous fraction are geraniol and terpineol. More than 200 components of the hop oil have now been isolated.

The least hop oil was found in fine hops (0.2–0.3 per cent), but in so-called crude hops there may be up to four times as much. Hop oil is practically insoluble in water but it is high volatile in water vapour. Therefore, it is not involved in the production of beer during the period when fresh hops are processed. More than 90 per cent of the oil volatilizes during the boiling period and the traces found in the hopped wort disappear during the primary and secondary fermentation. Therefore, no hop essential oils can be found in the finished beer.

The essential oils are transformed by oxidation and, consequently, their chemical and physical properties are changed. In particular their solubility increases and their volatility decreases. Under certain conditions some of the products of this transformation can pass into the finished beer and reduce its quality. Such an undesirable effect is likely when hops of a particular age are processed.

It can thus be concluded that the hop essential oil is useful in commerce only when the group to which a particular of hop belongs might be distinguished by scent.

Other constituents

In addition to resins, tannins and oils many other constituents are present in hop, such as saccharides, nitrogen compounds, lipids, waxes, sulphur dioxide and heavy metals. These components have no essential effect on the technology of production nor on the quality of the beer. The heavy metals (Cu, Zn, Mn, Fe) are taken up by the hop plant mostly from the chemicals used as crop protectant sprays. These are normally adsorbed onto the bitter dregs during the hopped wort production. Sulphur dioxide is not present in the original hops but arises from the treatment of the crop with sulphur. BRENNER and BERNSTEIN (1975) stated that the frothing quality of beer, more particularly, the stability of the froth will be adversely affected if the content of sulphur dioxide in the hopped wort exceeds 25 mg per litre. A greater content of nitrogen compounds will unfavourably affect the activity of the yeast. According to POSTL (1975) certain microorganisms are able to reduce nitrates to nitrites which are toxic to yeasts. When very old hops are processed their lipids can spoil the flavour of the beer. Nitrogen compounds consisting mainly of proteins and their decomposition product do not themselves affect the quality of beer, but according to NARZISS and BELLMER (1975), their reaction with saccharides produces substances which intensify the colour of beer. Pectins increase the effect of bitter substances.

This section has included a brief description of the typical features of substances contained in the hop as well as an indication of their effect on the technology and quality of beer. This survey shows that the dominant components are the hop resins. These substances are, however, unstable so that undesirable changes occur during storage and these changes gradually reduce the value of the hop for brewing. Even under good storage conditions the brewing value of hop decreases by 10–15 per cent per year.

Under the classical brewing process there is a poor level of utilization of the bitter substances from the hop. Thus, experience with production of Czech beers hopped with large doses, approximately 85 per cent of the bitter substances pass into the wort, i.e. approximately 15 per cent of the hop resins remain in the hops. Taking the quantity of bitter substances in hopped wort as 100 per cent, then the other steps in the process cause such great losses that their quantity in the finished beer is only 20–25 per cent. Most of the loss of bitter substances occurs in the cooling of hopped wort (due to the adsorption of the bitter substances on the separated dregs) and during the main fermentation due to changes of pH value, adsorption on yeasts, and the removal of bitter substances in foam layer. Further losses of bitter substances occur during the secondary fermentation and in the filtration of beer.

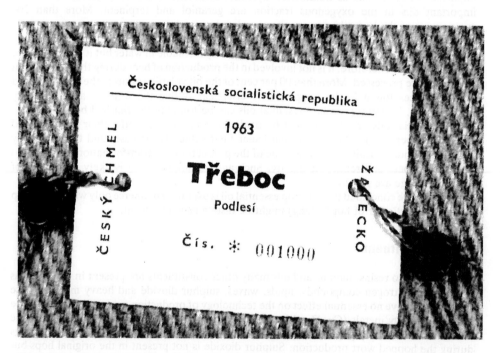

Fig. 1. Hop tokens: a – marking of bales before further procedures.

b – certificate of a consignment from Žatec region.

The fall in brewing value during hop storage and heavy losses of bitter substances in the production of beer are economically unsatisfactory. Therefore these questions are studied worldwide. They were very intensively investigated especially after World War II. The results of such research work provided numerous solutions which have been widely adopted in the brewing industry.

THE PRODUCTS OBTAINED FROM HOP CONES

Various modifications of hop cones have been used in breweries in many countries. Such modifications include hop extracts and ground or granulated hop cones. Hop extracts were used to a small extent in Czechoslovakia as early as 1960, but the use of ground or granulated hop cones did not occur in Czechoslovak breweries until 1970.

Hop extracts

There are two main types of hop extracts:
a) hop resin extracts (one stage type);
b) standard hop extracts (two stage type).
The group of two stage extracts have different ratios of resin to water.
One stage hop extracts consist almost solely of hop resins. The principle of one stage extraction involves the extraction of hop cones by a suitable organic solvent such as ethyl ether, methyl alcohol or methylene dichloride. After the completion of extraction the organic solvent is removed under vacuum and the resultant resins are filled into metal containers.

This type of hop extract was tried out in Czechoslovak breweries where it was shown that it can replace up to 20 per cent of hop cones without changing the taste of beer. The extract does not contain tannin, and therefore, it did not find wide application in Czechoslovakia. However, it is used in many foreign breweries.

The Research Institute of the Brewing and Malting Industry in Prague developed its own

17

technology to produce t w o s t a g e e x t r a c t s of the standard type. In the first stage of this process the hop is extracted by a suitable organic solvent as in the production of one stage extracts. Other substances in hop are then leached by hot water. This water fraction contains water-soluble substances, particularly the hop tannin. After the vacuum distillation of the organic solvent (first stage of extraction) and the concentration of the water fraction (second stage of extraction) both components are thoroughly homogenized, to be stored in metal drums. The two stage hop extract is produced in Czechoslovakia by the manufacturers Aroma in Nižbor.

Hop extract made by two stage extraction contains practically all the substances important in brewing, and it can, therefore, be used to replace a large proportion doses of the normal quantity of hop cones. Based on the known importance of individual fractions of hop resins in producing the bitterness of beer, a method was developed to determine the efficiency of hop extracts. As a result two sorts of two stage extracts are produced in Czechoslovakia; one with efficiency 1:5 (1 kg of extract replaces 5 kg of hop), and the other with efficiency 1:6 (1 kg of extract replaces 6 kg of hop). The main criterion of efficiency is the content of soft resins, as determined by a modified Wöllmer's method. The standard specifies that extracts with efficiency 1:5 must have 27.5 ± 1.5 per cent of soft resins and for the 1:6 extract the content should be 32.5 ± 1.5 per cent. These values have been confirmed by large scale investigations in different breweries. These two levels of efficiency are valid for replacement by the extract of up to 70 per cent of hop. If more than 70 per cent is replaced, the efficiency decreases, so that with the total replacement of hop by extract its efficiency drops from an original 1:5 to 1:4.5 and from the original 1:6 to 1:5.5. The reason for this is that the combination of hop cones with extract leads to a better utilization of the bitter substances than when the hop is processed alone. When the hop is replaced totally by extract this positive factor disappears and the efficiency of the extract consequently decreases.

As compared with hop cones, *the two stage extract has many advantages which can be summarized as follows:*

1. The utilization rate of bitter substances from the extract is 20–25 per cent higher. This is due to the partial transformation of original bitter acids during the extraction process and also an immediate reaction of resins in the extract with the hot malt wort. Smaller losses during the primary and secondary fermentations are also important.

2. The individual resinous fractions in the extract are very stable.

3. The constancy of composition of hop extracts facilitates both the analytical assessment and the dosing based on it. An analysis certificate accompanies every extract delivered to the brewery.

4. The losses of desirable substances in the hopped wort are smaller when hop extracts are used and the quality of work invested in the removal of brewer's grain decreases.

5. The space required for the transport and storage of hop extracts is only 1/25 that required for baled hop cones.

6. A high economic return is reflected in the lower cost of producing of beer without any reduction in its quality.

These advantages of hop extracts have been confirmed and used in Czechoslovakia for more than 25 years, and there is ample scope for their continued application.

Ground and granulated hops

These forms of processed hops came into use in certain countries about 1965. The production consists principally in cleaning hop cones and their drying down to 5–6 per cent water content. The hops are then ground, homogenized and filled into packages under an inert gas. After this reduction in volume the ground hops may also be formed into granules.

Recently the production of ground hop was introduced into Czechoslovakia by the manufacturers Chmelařství in Žatec. The grinding mills were imported, but the granulation is made on Czechoslovak patented equipment. The Research Institute of the Brewing and Malting Industry ran performance tests in 1974 with ground and granulated hops from one set of hop cones, to determine whether grinding or granulation causes undesirable changes. The results of these analyses are in Table 1.

TABLE 1 Chemical analysis of hop cones (mean values)

Constituents	Cones hop	Ground hop	Granulated
	percentage content		
Water	6.9	5.2	4.0
	in dry matter		
Total resins	17.6	17.2	16.7
Soft resins	15.1	14.8	14.3
Hard resins	2.5	2.4	2.4
α-bitter acids	6.8	6.3	6.2
β-fraction	8.3	8.5	8.1
Tannin	4.6	4.6	4.6
Active bitter substances	15.6	15.6	15.2
	in total resins		
Soft resins	85.8	86.0	85.6
Hard resins	14.2	14.0	14.4
α-bitter acids	38.6	36.6	37.1
β-fraction	47.2	49.4	48.5

The table shows that the processing of hop cones lowers the water content. This is a positive factor because it aids in the stability of the resins. The content of total and soft resins, and α-bitter acids was slightly decreased. This probably results from mechanical losses, because the total amount of resins and other components should in theory increase in proportion to the water decrease. The amount of tannin did not change. The results of analyses of the components as relative percentages show that the differences between them are within the range of analytical errors. Bitterness expressed as active bitter substances is practically the same for the ground and granulated hop.

Commercial-scale performance tests lead to following conclusions:

1. The utilization rate of bitter substances is, on average, about 10 per cent higher from ground and granulated hop than from hop cones providing a centrifuge tank is installed to cool the hopped wort and to separate the bitter dregs. In the classical process, using a filter press the rate drops to 5 per cent and in addition the larger quantity of bitter dregs makes the work with filter press difficult.

2. Ground and granulated hop improves the function of the centrifuge tank because the sedimentation of the bitter dregs is accelerated and the cone of sediment at the bottom of the tank is firmer.

3. No difference between the experimental and control beers have been detected by tasting panels.

4. The stability of resins in ground and granulated hop, packaged under inert gas, appears to be practically unlimited.

5. As with hop extracts the storage conditions and space required are less demanding than for baled hop cones.

These results of our investigations are similar to those published in scientific periodicals abroad. The final economic effect, which depends on the price of ground and granulated hop is at present less favourable than that obtained with hop extract. It may be assumed that the situation will change and that the consumption of processed hops will increase. This could depend on the rate at which centrifuge tanks can be installed.

Manufacturers of ground and granulated hops can offer their products with a constant composition and this makes the calculation of the amounts required for the production of beer very easy. This situation, however, requires a systematic analytical check, not only of processed hops, but also of the finished products. Providing that ground and granulated hops can be produced with constant composition, they will find application in breweries throughout the world.

There are other hop products available on the world market and in use in foreign breweries. Among them are the so-called enriched hops and mixed products, but such preparations are not yet regulary produced in Czechoslovakia. Therefore, in the following paragraphs are results published in foreign scientific periodicals.

Enriched hop

Production of the enriched hop proceeds as follows: the hop cones are dried, cleaned and ground at a cryogenic temperature (-35 °C). At this temperature the lupuline is separated from the other components of hop. The separate components thus obtained can then be re-mixed in any derived ratio. The final product is hermetically sealed under an inert gas. Before packaging it can be granulated. The resin content depends on the ratio of components, and this, in turn changes the utilization rate of the final product. A slightly higher content of tannin was found in enriched hop compared with hop cones, the water content and the quantity of nitrogenous substances is about half, and the proportion of oil does not change.

The main features of this product are similar to those of ground hops but the better utilization rate it provides depends on the lupuline concentration. The storage space required for enriched hop is even smaller than that for ground hops.

Mixed preparation (ground hop plus hop extract)

The mixed preparation is produced by mixing ground hop with single stage resin extract in a particular ratio. Ground hop serves as the carrier and liquid resin extract is sprayed onto it. This product is further thoroughly homogenized and is eventually granulated. After cooling it is hermetically packaged under inert gas. The product is standardized with regard to its content of α-bitter acids. According to GANZLIN (1976) the utilization rate of bitter substances is better with this preparation than the simple sum of the separate utilization rates for ground hop and hop extract.

– – – – –

All of the hop products currently available for the brewing industry have been described here. All of them have the long-term stability, particularly of the hop resins, required for materials important in brewing. Compared with hop cones they provide a better utilization rate of bitter substances. Seen from this standpoint, hop extracts take first place, the mixed preparations (ground hop with hop extract) are second and ground and granulated or enriched hops are third. All these products are used in the classical hopping process i.e. they are added to malt wort in copper tanks and boiled for 90 to 120 minutes. There are losses of bitter substances during further processing including the cooling of the wort and during primary and secondary fermentation in spite of the use of these preparations, but such losses are smaller than those which occur with the processing of hop cones.

The aim of reducing further losses of bitter substances during the production of beer promoted the search for fully soluble preparations which can be used for so-called cold hopping. Such materials make it possible to add the required amount of bitter substances to the beer during last stages of the process. Recently two preparations for cold hopping have become available on the world market; these are the iso-extract and hulupone extract.

Iso-extract

Iso-extract is completely soluble in water. It contains only the transformed products of α-bitter acids, an emulsifying agent and water. Its content of iso-α- acids ranges from 20 to 50 per cent. The production is based on the extraction of hop or hop extract by hexane or petroleum ether to provide a fraction of soft resins. Then the pH is raised, suitable catalysts (calcium, magnesium and manganese salts) are added and the solution of resins is isomerized at high temperature. Thus, the α-bitter acids are transformed into the iso-α-acids which pass from the organic phase into the water phase. The isomerization is almost quantitative. The aqueous solution of iso-α-acids is separated from the organic phase and acidified so that the iso-α-acids can be taken up into an organic solvent. The relatively pure iso-α-acids thus produced are diluted with alkaline solution and after adding emulsifying and froth-limiting agents, are filled into metal containers. As mentioned above the losses of bitter substances are minimized if this iso-extract is added during final stages of beer production.

At present the use of iso-extract in Czechoslovak breweries is limited. Pilot plants have been built and in large test runs the total hopping is lowered by 25 per cent. The required bitter substances, in the form of iso-extract, are added after the filtration of the beer. The iso-extract for these test runs is being imported. If the results indicate that the quality of beers is maintained, the use of iso-extract would seem to indicate a means of reducing costs.

The use of iso-extract shows greater promise in the production of new types of beer, such as dia-beer (i.e. beer for diabetics) where the losses of bitter substances are normally greater than during normal beer production. It is possible that, in this case, at least 50 per cent of the total hopping could be replaced by iso-extract. PFENNINGER and SCHUR (1972) found iso-extract being successfully used in 17 Swiss breweries. They also found that when iso-extract was used the utilization rate of iso-α-acids reached 80 per cent.

A by-product of the production of iso-extract is the socalled basic extract. This is a complex of soft resins from which the α-bitter acids have been separated. *The fraction thus contains β-bitter acids and non-specified soft resins and can be utilized in two ways:*

  a) for the production of hulupone-extract;
  b) for direct use in the brewing process.

Hulupone-extract

The manufacturing process for this material involves the addition of alkaline solution and hexane to the basic extract. The β-bitter acids pass to the alkaline solution and become oxidized. After acidification and extraction by hexane the solvent is removed, under reduced pressure to provide hulupone extract. The final product is modified in the same way as iso-extract, i.e. an emulsifying agent and water are added to provide the required concentration.

The original β-bitter acids are not bitter, and therefore, their utilization as hulupone extract represents an important contribution to the rationalization of the boiling procedure.

The hulupone extract could possibly be used with advantage in a mixture with iso-extract. According to PFENNINGER and SCHUR (1975) the bitterness of beers hopped with such preparations has the same quality as beer hopped with hop cones. We have not yet used hulupone extract but it seems likely that it would be more powerful than iso-extract in providing bitterness as indicated by the results of investigations on the transformation of β-bitter acids by SALAČ and DYR (1944).

Basic extract

Basic extract, a by-product of iso-extract, can be utilized in the classical hopping method. Because the basic extract contains only a resinous fraction, it shows considerable promise for the production of Czech beers. This application rate can reach 25–30 per cent. *The first test runs in a pilot plant gave the following information:*

a) yield of the basic extract, compared with normal Czech hops can be expressed by ratio 1:5; there is used the combination consisting of 75 per cent of hop cones and 25 per cent of basic extract;

b) the flavour of experimental beer was indistinguishable from that of control beer.

Recent investigations have concentrated on the detailed analysis of the preparation with a view to its eventual utilization as a standard extract. It can be concluded that the basic extract will be found a place in the production of Czech beers.

TABLE 2

Constituents (percentage in dry matter)	European climbing hop		
	stems	leaves	waste leaves
Total nitrogenous substances	5.87	15.75	14.35
Ether extract	1.22	5.76	5.41
Crude fibre	38.60	11.25	11.84
Extracted nitrogen-free substances	48.74	49.18	48.92
Digestible nitrogenous substances	2.47	9.16	8.26
Ash	5.57	18.06	19.48
Found in ash:			
Over 10 %	Ca, Mg, K	Ca, K	Ca, K
From 1 to 10 %	P	Mg, P	Mg, P
From 0,1 to 1 %	Si	–	–
From 0,01 to 0,1 %	Al, B, Zn	Si, Al, B, Zn	Zn, Si, Al, B
Less than 0,01 %	Sr, Mn, Cu Pb, Fe, Ba Cr, Na	Sr, Mn, Cu Pb, Fe, Na Ba, Cr, Mo	Mn, Sr, Na Cu, Fe, Pb Ba, Cr, Mo
Ash insoluble in hydrochloric acid (%)	0.56	4.83	6.48
Carotene (mg . kg^{-1})	–	69	63
Note: waste leaves = leaves after machine harvest.			

To complete the list of possible agents for replacing hop cones the s y n t h e t i c b i t t e r s u b s t a n c e s and s y n t h e t i c i s o - α - a c i d s must be included. Such a preparation was produced in small quantity by the Swiss company Givaudan in Zürich. The substance is likely to find the same application as the iso-extract made from hop cones. Preliminary pilot plant tests made by PFENNINGER and SCHUR (1975) showed that the preparation has an almost 100 per cent utilization rate and that there was no reduction in the qualiy of the beer. As yet there has been no economic comparison with the classical boiling process and the costs of synthetically produced iso-α-acids are probably higher than those obtained from hops. The fact no production line for the synthetic material was ever built seems to bear out this latter suggestion.

Substances for use in cold hopping have not generally found such wide application as those used in the classical brewing method. Nevertheless the further introduction of these products into brewing must be considered, because iso-extract and hulupone-extract are stable materials with high utilization rates and could help reduce the losses of bitter substances during the production of beer.

– – – – –

This section has discussed recent knowledge concerning the application of hop cones and different hop preparations on the production of beer. The properties of the substances have been related to the production of typical Czech beers. Several hop preparations satisfy requirements for their application to beer production in Czech breweries. They can replace a large proportion of the hop cones normally used in the classical hopping method. This is particularly true of the standard hop extract, of ground and granulated hops and even of the

Analyses of stems and leaves with regard to nutritive values

Japanese hop		Great nettle	
stems	leaves	stems	leaves
9.75	18.58	8.56	21.72
1.26	2.54	0.65	1.56
28.10	8.69	46.75	8.16
51.94	45.21	33.31	47.32
5.23	11.03	6.02	15.30
8.95	24.98	7.37	21.24
Ca, K, Mg	Ca, K	Ca, Mg, K	Ca, K
P	Mg, P	P, Si	Mg, P
Al, Si	Si, Al	Al	–
B, Mn, Zn	B, Zn	Mn, Zn, B, Si	Mn, Zn, Al, Si, B
Ba, Sr, Fe,	Ba, Cr, Na,	Se, Na, Cu,	Ba, Sr, Na,
Cr, Na, Cu,	Mn, Cu, Pl,	Pb, Fe, Ti,	Cu, Fe, Pb
Pb, Ti	Fe, Ti, Sr	Mo	
2.09	7.63	0.87	7.90
–	55	–	44

combined materials such as ground hop plus extract. The resinous (one stage) extract or basic extract can be used to only a limited extent in the classical boiling procedure. There will be a similar limitation in the eventual application of substances for cold hopping (iso-extract, hulupone-extract) in Czech production. In comparison with hop cones these preparations have many advantages two of which are dominant, namely their stability and their higher utilization rate as bitter substances. These factors have recently affected the trends in the consumption of hop cones and hop preparations. According to information provided by the Barth company roughly 50 per cent of the global harvest of hops is now processed as hop extracts, and ground and granulated hops. According to HUGO (1976) the use of hop cones has been almost abandoned in West European breweries, only 10–15 per cent of which now use hop cones exclusively, for beer production. In the future there is likely to be further replacement of hop cones by different hop preparations as a contribution to lowering the cost of beer production. The Czechoslovak brewing industry is taking account of this trend, and except for certain top breweries which produce for export Czech breweries now mainly use standard hop extract combined with hop cones in the ratio 1:1. For the future it is foreseen that present rate of hop extract will continue to be used while at the same time hop cones will be replaced by ground and granulated hop. The application of suitable preparations for cold hopping are likely to be used, but on a small scale.

The present situation together with the expected developments in breweries worldwide will put greater demands on research, breeding and trade. The production of hops has to be so organized as to satisfy all the requirements of the breweries. Bearing in mind the traditions of brewing and the typical properties of hops it is clear that fine aromatic hops with a typical combination of particular components will continue to be planted. It is, however, essential that the desirable components are represented to the greatest possible extent. This should be ensured in Czechoslovakia by the new state standards concerning the evaluation of hops. These specify the determination of bitterness rate by an objective analytical method so that an accurate state purchasing price can be decided. This should guarantee that the crop will be picked at the optimal time of maturity. Such hops will serve well as the raw material for the production of different preparations to be applied in the classical hopping method.

In view of the global trends in the consumption of hop and its derivatives a good hop producer must be aware of forthcoming changes, and those making hop preparations from homeproduced hops must preserve their acknowledged valuable properties. This being the case Czech hop products of constant composition will continue to be offered and will maintain their leading position in world markets as has been the case in the past with Czech hop cones.

UTILIZATION OF SECONDARY PRODUCTS

Secondary products from hops include leaves and stems. The results of their analyses as fodder are shown in table 2. For comparison, similar analyses of Japanese hop (*Humulus japonicus* Sieb. et Zucc.) and great nettle (*Urtica dioica* L.) are also presented. The leaves of hop are rich in proteins, mineral substances, vitamins, carotene and are low in crude fibre. Therefore they provide an excellent fodder when fresh or after conservation, and especially after drying.

The feeding value of stems and shoots is essentially lower. Because they often include fragments of wires and hooks they are unsuitable for feeding and can be used only as an organic mass for composting. They enrich composts not only as an organic mass but also provide important macroelements and microelements.

CHAPTER II
BIOLOGICAL BASIS OF HOP PRODUCTION

As in other branches of plant production the biological, technical and economical aspects are each involved in the technology of hop production.

The biological properties of hop form the basis of the technology of production. This however does not mean that the technology is entirely dominated by these properties. Any non-adherence to technical and economic principles is reflected in the efficiency of production, but non-adherence to basic biological principles endangers the very essence of production. Thus a knowledge of the biology of hop plants is absolutely necessary for the scientific management of the technology of production. Likewise knowledge of the biological processes in the soil is needed as they are involved in plant nutrition, as is also knowledge of the biology of weeds, and the diseases and pests of the hop.

Genetic properties which have arisen and become fixed over a l o n g h i s t o r i c a l d e v e l o p m e n t (p h y l o g e n e s i s) are reflected in the biology of hop plants. The process of i n d i v i d u a l d e v e l o p m e n t (o n t o g e n e s i s) concerns the life of the individual plant during its different periods, and the ultimate yield. The properties of particular organs, including the harvested product i.e. hop cones, a r e a f f e c t e d b y t h e i r i n d i v i d u a l o r i g i n , d e v e l o p m e n t a n d f o r m a t i o n , i . e . b y o r g a - n o g e n e s i s d u r i n g e a c h o n e y e a r l i f e c y c l e o f t h e h o p p l a n t . Organogenesis is directly involved in the quantity of the crop and in profit. Organogenesis cannot be fully understood without a knowledge of the ontogenesis and phylogenesis of the hop plant.

Perennial plants of the European hop *(Humulus lupulus* L. subspecies *europaeus* Ryb.) (mentioned hereafter as hop) have many biological properties of agricultural and economic importance which affect the yield and quality of the main product – hop cones. From the agricultural point of view it is very important that there is little fluctuation in yield and quality over a long period.

Knowledge of the morphology and physiology of the whole hop plant and its individual organs is essential if the maximum yield and quality of hop cones are to be achieved. This approach offers a better understanding of the internal regulation of the plant and its organs, as well as the external relations between hop plant and its environment. It is not feasible, in one chapter, to consider all of the biological properties of the hop plant so far discovered. Therefore the factors to be discussed here are those which affect the utilization of the hop plant. Some parts of hop plant, and the properties they possess have different names in botanical terminology from those used in hop production. The origin of this technical terminology has a very interesting history.

According to historical documents the hop has been cultivated in Czechoslovakia for nearly a thousand years. The culture in Czechoslovak countries reached such a high professional level, as early as the Middle Ages, that the hop growers, sometimes together with wine growers, formed themselves into independent guilds. Thus, the Czech hop-growing terminology, sometimes similar to that used in wine-making came into existence and has been traditionally preserved to the present day. Where the terminology differs both the botanical and hop-growing terms will be quoted here.

BIOLOGICAL FEATURES OF HOP PLANTS

The European climbing hop is a perennial plant, above-ground organs of which die back every year at the onset of the cold period. Only well-grown underground organs and roots with secondary thickening survive. The *perennation of the hop plant* relies on the ability of underground dormant buds to survive for up to four years. All buds older than four years, active as well as dormant, decay and are replaced annually with new dormant buds. This complement of dormant buds maintains the hop plant over many years. Some organs of the underground stem system, however, survive for more than five years. Occasionally atrophied tissues become putrescent and diseased. These infect healthy tissues thus lowering the vitality and productivity of hop plant and causing its decay. Vacant places in hop gardens are planted with new roots so that the overall productivity and average age of the stock is improved. Wild hops, which may propagate vegetatively or by seed, can survive on one site for an indefinite period, but the stock of hops cultivated in gardens, where seed propagation is not possible, is entirely dependent on the systematic replacement of decayed individuals so as to maintain and improve the stock.

Four organ systems can be distinguished on compact hop plants; two of them are underground (roots and modified stems) and two above-ground (vegetative organs and generative organs). The *main structures and properties of individual systems* are shown in Tables 3 and 4.

All analyses of cones were done at what was regarded as the time of their technical maturity. This time is that at which the cones would be regarded as ready for picking. The analyses in the tables were done on two well-grown five-year old vines of 'Osvald's clone No. 72'.

The dry matter percentage during the picking period, of the individual systems was as follows: vegetative system – 44 per cent, generative system (hop cones) – 22 per cent, root system – 19 per cent and underground stem – 15 per cent. These proportions would be different at different times of the annual cycle. Moreover the ratios of carbon to nitrogen and especially of calcium to nitrogen change in particular organ systems and serve as simple indicators of the biological age of individual organs and systems.

A comparison of the main groups of substances in the different systems of hop plant show the content of crude fibre to be the most constant ranging from 15.72 per cent in the dry matter in hop cones to 27.21 per cent in the dry matter in root system. In contrast, the ether-extract is the most variable ranging from 0.81 per cent in the underground stem system up to 18.96 per

TABLE 3a **Analysis of dry matter of five-year old hop plants**

Plant part	Constituents (percentage of dry matter)					
	ash	nitro-genous substan-ces	ether extract	crude fibre	nitrogen free extractive substances	total organic substances
System of underground stem organs:						
Rootstock	6.68	9.93	1.00	26.37	56.02	93.32
Root system	5.78	7.88	0.81	27.21	58.32	94.22
Underground part	6.18	8.78	0.90	26.84	57.30	93.82
System of above-ground vegetative organs:	12.82	9.36	2.76	26.75	48.31	87.18
Cones	8.60	20.68	18.96	15.72	36.04	91.40
Above-ground part	11.42	13.11	8.13	23.10	44.24	88.58
Whole plant	9.63	11.63	5.66	24.38	48.70	90.37

TABLE 3b **Analysis of dry matter of five-year old hop plants**

Plant part	Constituents (percentage of dry matter)					
	starch	reducing sugars	other extractive nitrogen free substances	total nitrogen free extractive substances	glucose from starch	total monosa-charides
System of underground stem organs:						
Rootstock	3.71	1.98	50.33	56.02	3.92	5.90
Root system	5.84	2.70	49.78	58.32	6.48	9.18
Underground part	4.90	2.38	50.02	57.30	5.35	7.73
System of above-ground vegetative organs:	1.10	4.11	43.10	48.31	1.22	5.33
Cones	0.03	3.00	33.01	36.04	3.33	6.33
Above-ground part	0.75	3.74	39.75	44.24	1.92	5.66
Whole plant	2.17	3.28	43.25	48.70	3.09	6.37

TABLE 4a **Dry matter content and major elements in dry matter of five-year old hop plants**

Plant part	Dry matter		Dry matter content			
	total (g)	ratio (%)	C (%)	N (%)	C/N (C : N)	10 Ca/N (Ca : N)
System of underground stem organs:						
Rootstock	115.00	15.05	46.44	1.588	29.666	8.245
Root system	145.75	19.08	47.13	1.260	37.405	11.960
Underground part	260.75	34.13	46.92	1.405	33.992˙	10.322
System of above-ground vegetative organs:	336.60	44.05	43.587	1.498	37.899	19.102
Cones	166.70	21.82	45.495	3.310	13.516	3.891
Above-ground part	503.30	65.87	44.219	2.098	29.823	14.064
Whole plant	764.05	100.00	45.141	1.861	31.246	12.787

TABLE 4b **Dry matter content and major elements in dry matter of five-year old hop plants**

Plant part	Dry matter content						
	Cl (%)	S (%)	P (%)	Ca (%)	K (%)	Mg (%)	Fe (%)
System of underground stem organs:							
Rootstock	–	0.140	0.343	1.289	1.405	0.198	0.155
Root system	0.033	0.146	0.277	1.496	1.111	0.186	0.171
Underground part	0.033	0.143	0.306	1.405	1.241	0.191	0.164
System of above-ground vegetative organs:	0.465	0.191	0.285	2.778	1.467	0.283	0.058
Cones	0.230	0.341	0.564	1.288	3.274	0.247	0.068
Above-ground part	0.387	0.241	0.377	1.505	2.066	0.291	0.061
Whole plant	0.270	0.208	0.353	1.471	1.784	0.244	0.096

cent in cones. A similar contrast is found in the content of nitrogenous substances ranging from 7.88 per cent in the root system up to 20.68 per cent in cones. Likewise the nitrogen-free substances range from 58.32 per cent in the root system up to 36.04 per cent in cones. There was little overall contrast between contents of organic substances, but ash (mineral substances) ranged from 12.82 per cent in the above ground vegetative system up to 5.78 per cent in the underground root system.

Underground parts of hop plant, their function and structure

The two underground organ systems of hop plants are different in their morphology and main functions. These are the underground stem organs, i.e. the so-called rootstock, and the root system.

T h e u n d e r g r o u n d s t e m o r g a n s are of primary importance not only because they are situated between the root system and above-ground parts but, in particular, for the key function of their dormant buds. These, as previously mentioned, remain functional for a period up to four years and so provide a basis for the perennial life of the hop plant.

T h e r o o t s y s t e m includes all roots, regardless of the underground part of the stem from which they develop. The primary function of roots is to obtain water and nutrients from the soil, to ensure that these are transported to other organs and to undertake certain primary transformations. Roots, and especially root tubers, serve as a storage organ.

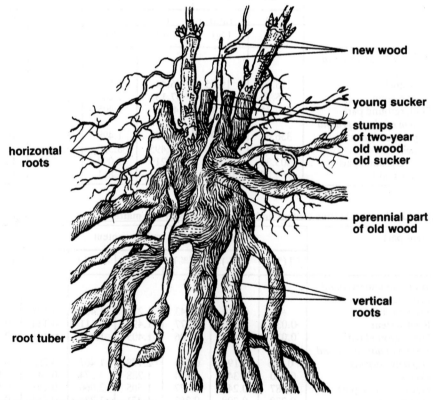

Fig. 2. Underground part of hop plant.

As is common in dicotyledonous plants there are essential differences in anatomical structure between the two systems (see Fig. 2). The difference in function between the two underground systems is reflected in their different structures as well as by differences in chemical composition of different biological ages of the rootstock and root system as shown in tables 3 and 4. These indicate that the root system of a five-year old plant has more dry matter and higher biological age than has the rootstock. The root system also contains more stored matter, especially starch, whereas the rootstock contains more ash and total nitrogenous substances.

System of underground stem organs-rootstock

The rootstock of the hop plant includes all of the organs produced by modification of the stem below soil level. The different environment wherein they grow and their changed functions are reflected in differences in their structure as compared with the above-ground stem. The basic morphology and anatomy of the above-ground stem and underground rootstock have essentially similar features, and it is only in proportions that the individual parts and tissues differ. Internodes of the underground stem are much shortened and thicker. Primary xylem bundles are found at the centre of a section of internode. The secondary xylem is towards the outside consisting of many vessels with diameters up to 0.3 mm. A ring of secondary phloem, consisting of many vessels and parenchyma, is found outside of the secondary xylem. Secondary phloem is more developed in the underground stem than in its above-ground counterparts. This results in the thickening of the phloem which can thus serve as a storage tissue for the deposition of stored substances. In the autumn organic storage substances are also deposited in the xylem vessels which then cease to be a transport channel. They become filled with round plasmatic formations called thyllae.

Secondary bark (periderm), with visible lenticels appears on the underground stems. These are narrow structures a few millimeters long. They allow gas exchange, especially the inlet of air and the outlet of carbon dioxide produced in and tissue respiration.

As well as the *buds* the underground stem system has *vertical stem organs (rhizomes)* which have the appearance of one year old new wood or of perennial old wood according to their age. *The horizontal stem organs,* known as *suckers (stolons)* also appear as *one year old structures* or as *perennial and old.* There are more changes in the structures of underground stem organs than can be seen as morphological modifications. It is, therefore, necessary to distinguish between *the one year old organs (young wood)* and *the older organs (well-grown wood).* The calendar age of the well-grown wood can be determined by its number of annual rings. Each year one annual ring 2–4 mm wide is produced with new secondary xylem and phloem as a result of the activity of cambium. In these circumstances the dormant buds retain their vitality in only the four youngest annual rings and are no longer viable in the fifth annual ring. However in the above-ground organs this vitality disappears much more rapidly, and here the dormant buds survive for only a couple of months.

The main function of the underground stem organs is to maintain the merismatic tissues during the unfavourable cold period. This function is very important, because it not only retains the facility for continuous growth of the underground system but it also ensures an early and rapid growth of the above-ground tissues at the start of the next vegetation period. In addition, cuttings taken from the underground stem organs can be for vegetative propagation.

Main physiological and biochemical differences between young and well-grown wood are clearly shown in Table 5 and 6. These tables indicate that young wood forms up to one third of the rootstock of five-year old plants. It contains more nitrogen and less calcium than well-grown wood so that its biological age, calculated according to the calcium-nitrogen index

(10 Ca/N), is essentially lower. Moreover, young wood contains less fibre but more extracted nitrogen, free substances and has more reducing sugars and total monosaccharides.

Their chemical composition indicates that the activated buds are considerably younger in the biological sense than the young wood whereas the well-grown wood is essentially older. The developing buds, however, represent only an insignificant part of the dry matter of the whole underground stem system so they have little effect on the overall chemical composition,

TABLE 5a   **Analysis of rootstocks of five-year old hop plants**

Plant part	Constituents (percentage of dry matter)					
	ash	nitro-genous sub-stances	ether extract	crude fibre	nitrogen free extrac-tive sub-stances	total organic sub-stances
New wood	6.05	8.89	1.00	23.12	60.94	93.95
New suckers	5.83	11.95	–	23.10	59.12	94.17
Young wood	5.94	10.37	1.06	23.11	59.52	94.06
Well-grown wood	7.02	9.73	0.97	27.86	54.42	92.98
Rootstock	6.68	9.93	1.00	26.37	56.02	93.32

TABLE 5b   **Analysis of rootstocks of five-year old hop plants**

Plant part	Constituents (percentage of dry matter)					
	starch	redu-cing sugars	other nitro-gen free extrac-tive substan-ces	total nitrogen free ex-tractive substan-ces	glucose from starch	total mono-sa-ccha-rides
New wood	6.90	2.58	51.40	60.94	7.66	10.24
New suckers	7.80	0.50	50.82	59.12	8.66	9.16
Young wood	7.45	1.57	51.00	59.52	8.14	9.71
Well-grown wood	2.02	2.17	50.23	54.42	2.14	4.41
Rootstock	3.71	1.98	50.33	56.02	3.92	6.06

TABLE 6a   **Dry matter content and major elements in dry matter of rootstock of five-year old hop plants**

Plant part	Dry matter		Dry matter content			
	total (g)	ratio (%)	C (%)	N (%)	C/N (C:N)	10 Ca/N (Ca:N)
New wood	18.6	16.17	46.98	1.422	33.038	8.439
New suckers	17.7	15.39	47.09	1.864	25.263	5.032
Young wood (subtotal)	36.3	31.56	47.03	1.636	29.247	6.778
Well-grown wood	78.7	68.44	46.49	1.557	29.859	8.921
Rootstock	115.0	100.00	46.66	1.588	29.666	8.245

TABLE 6b **Dry matter content and major elements in dry
matter of rootstock of five-year old hop plants**

Plant part	Dry matter content						
	Cl (%)	S (%)	P (%)	Ca (%)	K (%)	Mg (%)	Fe (%)
New wood		0.144	0.366	1.200	1.483	0.204	0.121
New suckers	0.036	0.181	0.498	0.938	2.210	0.213	0.117
Young wood (subtotal)		0.162	0.430	1.072	1.837	0.208	0.119
Well-grown wood		0.130	0.303	1.389	1.206	0.194	0.171
Rootstock		0.140	0.343	1.289	1.405	0.198	0.155

whereas the well-grown wood of five-year old plants makes up more than two thirds of the underground stem system and therefore has a highly significant effects on its chemical composition.

In contrast, in the physiological processes the situation is totally different because the activated buds, and the new suckers they form, are so physiologically active, that they affect the activity not only of the whole underground stem system but also that of other systems of hop plant.

Underground buds

Underground buds are either *dormant* or *active*. Dormant buds up to four years old are very small, and their detailed examination is, therefore very difficult. All underground and above-ground organs arise from these buds which are modified for different functions. This modification results either in a vertical or horizontal direction of growth. Horizontally oriented buds generate the suckers whilst the vertically growing buds generate new wood and above-ground stems.

New wood and new suckers have only one year dormant buds, which become active very early; usually in the case of the topmost buds, before cones are ready for picking. At the time of technical maturity of cones there is an active bud on every sucker and more on a new wood. Analyses of buds from five-year old hop plants are shown in Tables 12 and 13. These indicate that the excited buds have a very low biological age, but they are high in total nitrogenous substances and ash, which are important for the growth of tissues. Supporting substances such as fibre and calcium are less in quantity.

Young wood

Young wood comprises the underground stem organs up to one year old. It constitutes that underground part of stem known as new wood and the underground horizontal runners (suckers). The main function of this wood rests in the fact that new above-ground stems grow from its buds in the next growing season. Both parts of young wood can provide material for the vegetative propagation of the hop plant once they are separated from the mother plant. Both parts have the same essential functions but exist under different conditions, particularly in the direction of their growth. This affects the length of internodes as well as the morphological and anatomic structure of both organs.

N e w   w o o d grows from the old wood, vertically, towards the surface of the soil. Its internodes are short and very much thickened. Tillering nodes, with buds, are close together, especially in the basal part of the new wood. The development of new stems is thus graduated only in a vertical direction and there are only small spaces between the developing shoots. The length of new wood is that distance between the bud from which it develops on the old wood and the surface of the soil. Under natural conditions new wood would rarely be separated from the mother plant except in the case of atrophied or broken old wood. Therefore the main function of new wood in the wild hop is as a source of buds for the renewal of stems in the next season and this function is prevalent over that of vegetative propagation. In the cultivated hop both functions of new wood are used: in production stocks the function of renewal of stems is maintained, and among mother stocks the production of cuttings is used in vegetative propagation.

Certain buds on the old wood or old suckers produce n e w   s u c k e r s . These underground runners have many tillering nodes with dormant buds in the axils of scales. An active

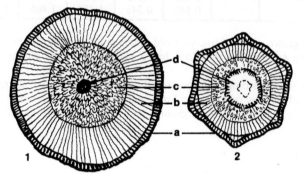

Fig. 3. Cross section of new wood (1) and above-ground stem (2): a – epidermis, b - bark, c – vascular ring, d – heartwood.

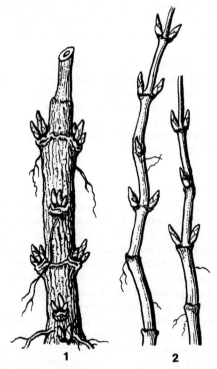

Fig. 4. New wood (1) and suckers (2).

32

top bud has very strong apical dominance which inhibits the development of dormant buds lower down the sucker. If this dominance is disturbed the sucker will become branched. These suckers grow horizontally and have longer internodes so that they can be over a metre in length. They are usually thinner and produce more adventive roots. In cross section less pith and cortex are found than in the new wood whereas the conducting tissues are well developed.

In addition to the morphological differences, certain differences in chemical composition of suckers and new wood were found in five-year old plants, as shown in Tables 5 and 6. These tables indicate that the suckers contain more nitrogen and less calcium so that their calcium to nitrogen index is lower than that of new wood.

Cuttings obtained from suckers, because of their physiological and biochemical properties, form better planting material than cuttings from new wood, and should be used for cultivation of rooted hop cuttings. However, because of their structure they wither more quickly than cuttings from the new wood and, therefore, they need better care during storage and transport.

Well-grown wood

Well-grown wood includes o l d w o o d and o l d s u c k e r s , arising from new wood, i.e. one year old tissues.

Old suckers of normal plants are, however, at least two years younger than old wood. The specific differences between these two tissues are similar to those between new wood and new suckers. The calendar age of old wood and old suckers can be established by the number od annual rings. Cambial activity produces one annual ring 2–6 mm thick, each year, in old wood and a thinner ring in old suckers. Well-grown wood has a greater biological age than young wood as can be seen from Table 6. Well-grown wood also has a greater content of ash and fibre and less nitrogenous substances and extracted nitrogen free substances as shown in Table 5.

Fig. 5. Old wood with bases of roots.

Root system

The root system includes all roots without regard to the underground organ which produces them. All roots can be morphologically distinguished from the underground stem as they possess no bud nodes and their anatomical structure is different (see Fig. 6). Primary roots

33

normally have a diarchic, and sometimes a triarchic, vascular bundle (with two medullary and two phloem rays). Cambial tissue produces vigorous secondary tissues in secondary thickened roots with xylem as a compact cylinder with numerous vessels. The secondary phloem is also robust with parenchyma used for the storage of reserve substances. Secondary bark is also developed with many lenticels and some of the cells of the phloem and secondary bark contain resins and polyphenolic substances.

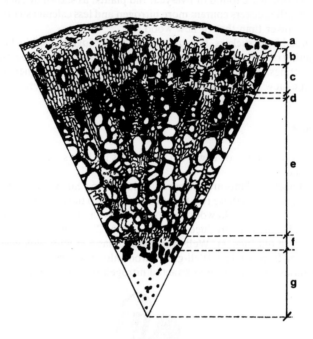

Fig. 6. Sector of a cross section of root: a – epidermis, b – bark, cortex, c – phloem, leptome, d – cambial ring, cambium, e – vascular cylinder, hadrome, f – endodermis, g – heartwood.

Two basic categories of roots can be distinguished according to their maturity. First group consists of *skeletal roots*, which includes all the secondary thickened roots forming the skeleton of the whole root system. The second group consists of *terminal, active rootlets* (diameter c. 1 mm) including the youngest rootlets with a primary anatomy structure and those just starting their secondary thickening procedure. The two groups have different basic functions. The terminal rootlets, through their root hairs take up water and solutions of mineral substances and they participate in primary metabolism including the conversion of inorganic nitrogenous substances into organic. The *skeletal roots,* both *vertical* and *horizontal,* facilitate the ascent and descent of streams of plant liquors involved in the deposition of reserve substances. This storage function depends on special modifications of vertical roots and their branches, to form *root tubers,* which will be described under a separate heading.

Main physiological and biochemical differences between the various categories of roots are shown in Tables 7 and 8. The tables indicate that the root system of five-year old plants during the picking period consists mainly of vertical skeletal roots followed by horizontal skeletal roots, root tubers and, finally, terminal rootlets.

The biological age, according to the calcium to nitrogen ratio during the picking period, was highest for skeletal roots and lowest for root tubers.

34

With regard to the main groups of substances the highest content of ash and nitrogenous substances was found in terminal rootlets, but the highest content of starch, and consequently of all monosaccharides, was found in root tubers.

TABLE 7a Analysis of root system of five-year old hop plants

Root system	Constituents (percentage of dry matter)					
	ash	nitroge-nous substances	ether extract	crude fibre	nitrogen free extractive substances	total organic substances
Horizontal roots	5.28	7.14	1.02	27.41	58.05	94.72
Vertical roots	5.75	8.02	1.02	30.44	55.79	94.25
Skeletal roots (subtotal)	5.56	7.67	1.02	29.23	56.69	94.44
Root tubers	5.31	8.58	–	19.94	66.17	94.69
Terminal rootlets	11.09	9.63	–	11.03	68.25	89.91
Other roots (subtotal)	6.86	8.86	–	17.56	67.21	93.14
Roots in total	5.78	7.88	0.81	27.21	58.32	94.22

TABLE 7b Analysis of root system of five-year old hop plants

Root system	Constituents (percentage of dry matter)					
	starch	reducing sugars	other nitrogen free extractive substances	total nitrogen free extractive substances	glucose from starch	total mono-sacha-rides
Horizontal roots	6.50	3.60	47.95	58.05	7.22	10.82
Vertical roots	3.33	2.17	50.29	55.79	3.70	5.87
Skeletal roots (subtotal)	4.61	2.74	49.36	56.69	5.10	7.84
Root tubers	14.40	2.90	49.53	66.17	16.00	18.90
Terminal rootlets	4.50	1.46	62.29	68.25	5.00	6.46
Other roots (subtotal)	11.75	2.52	52.94	67.21	13.06	15.58
Roots in total	5.84	2.70	49.78	58.32	6.48	9.18

TABLE 8a Dry matter content and major elements in dry matter of root system of five-year old hop plants

Root system	Dry matter		Dry matter content			
	in total (g)	ratio (%)	C (%)	N (%)	C/N (C:N)	10 Ca/N (Ca : N)
Horizontal roots	48.00	32.93	47.360	1.142	41.470	13.00
Vertical roots	72.50	49.74	47.125	1.283	36.730	12.36
Skeletal roots (subtotal)	120.50	82.67	47.219	1.227	38.483	12.62
Root tubers	18.50	12.69	47.345	1.373	34.483	8.01
Terminal rootlets	6.75	4.64	44.955	1.541	29.173	11.06
Other roots (subtotal)	25.25	17.33	46.706	1.418	32.938	8.83
Roots in total	145.75	100.00	47.130	1.260	37.405	11.96

TABLE 8b **Dry matter content and major elements in dry matter of root system of five-years old hop plants**

Root system	Dry matter content						
	Cl (%)	S (%)	P (%)	Ca (%)	K (%)	Mg (%)	Fe (%)
Horizontal roots	0.032	0.154	0.317	1.484	1.186	0.187	0.161
Vertical roots	0.031	0.143	0.264	1.586	0.968	0.185	0.185
Skeletal roots (subtotal)	0.031	0.147	0.285	1.545	1.055	0.186	0.175
Root tubers	0.043	0.144	0.240	1.000	1.371	0.185	0.151
Terminal rootlets	–	–	–	1.704	1.394	–	–
Other roots (subtotal)	0.043	0.144	0.240	1.261	1.377	0.185	0.151
Roots in total	0.033	0.146	0.277	1.496	1.111	0.186	0.171

Terminal rootlets

The terminal rootlets, also known as active rootlets, are young rootlets with a primary anatomical structure without secondary thickening and rarely exceed 1 mm in diameter. Starting from the tip they have following parts: root cap meristem, zone of cell elongation and root hair zone.

The main function of terminal rootlets is to take up the soil solutions and hence to provide the plant with its supplies of water and nutrients. In addition, by their function in root elongation, they increase the size of the root system. The terminal rootlets also show morphological differences depending on their specialized function. Thus, they can be categorized according to their appearance and especially according to their thickness, into *thin end-rootlets* (0.2–0.6 mm) and *thick end-rootlets* (diameter about 1 mm). The precise criterion for these categories is, however, their histological structure. Thin end-rootlets have two protoxylem groups with a thin primary cortex but without a clearly defined epidermis (rhizodermis). Thick end-rootlets have a central cylinder of three protoxylem groups, a clearly defined parenchyma layer in the primary cortex and a well-differentiated epidermis.

The number of thick terminal rootlets is normally much smaller than that of the thin ones and in hop seedlings there are very few, one at the tip of the tap root and others at the tip of lateral roots. Branches of horizontal roots terminate with a large number of thin terminal rootlets.

The very different thickness of the two forms of terminal rootlets results from their adaptation to their main functions. The thick rootlets are responsible for growth and penetration into the soil in both the vertical and horizontal direction, whereas their rootlets are concerned with absorption of the soil solution. Their numbers and physiology reflect these functions.

In the past it was generally accepted that water uptake is proportional to the uptake of mineral nutrients. MEYER and ANDERSON (1952), however, discovered that most of the absorption of water occurs in the root hair zone, and mineral salts are mainly taken up in the zone of cellular elongation. They, therefore, concluded that the water mechanism is different from that of nutrients and that there is consequently no direct connection between water and nutrient uptake.

It was discovered (RYBÁČEK, 1978) that the terminal rootlets of the hop plant have *hydrotropism* as well as *chemotropism* which facilitates soil penetration and root branching in wetter soil layers.

At a certain time one or more layers of epidermal cells begin to suberize and lignify over the root-hair zone. Older cells also develop into corky layers. This intermittent zone differs in its

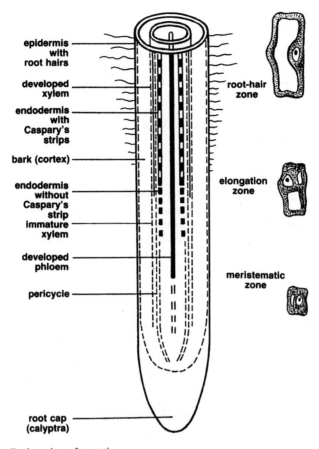

epidermis with root hairs

developed xylem

endodermis with Caspary's strips

bark (cortex)

endodermis without Caspary's strip immature xylem

developed phloem

pericycle

root-hair zone

elongation zone

meristematic zone

root cap (calyptra)

Fig. 7. Longitudinal section of root tip.

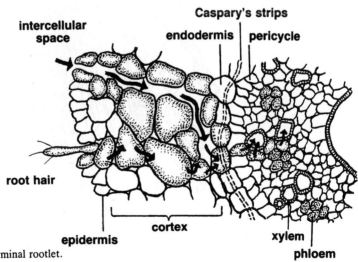

intercellular space

Caspary's strips

endodermis

pericycle

root hair

epidermis

cortex

xylem

phloem

Fig. 8. Cross section of terminal rootlet.

structure from that of normal terminal rootlets and therefore is usually included in the skeletal roots.

Both types of terminal rootlets are short-lived – a maximum of a few months, so they appear and disappear during the annual life cycle of the hop plant. When thin terminal rootlets become old they atrophy and fall off. Once the thick terminal rootlets have ceased to elongate the surface cellular layers becomes cutinized so their epidermis becomes firm and is protective against unfavourable conditions. Due to the change in their histology the thick terminal rootlets become similar to the skeletal roots. With the return of better conditions in spring, new thin terminal rootlets arise.

Skeletal roots

The skeletal roots constitute the basis of the whole root system. They include all secondary thickened roots, except the root tubers, which serve as storing organs. The main function of the skeletal roots is to anchor the plant in the soil and to distribute plant liquors. At the same time they consitute a skeleton, which at its extremities is surrounded by the terminal rootlets.

Every year the skeletal roots produce an annual ring and, if there is no limiting factor as e.g. a high ground water table, they become longer. Hence, the average length of an annual increment can be easily established according to the number of annual rings. But the establishment of total age according to the number of annual rings in the oldest part of the root may be not quite accurate, especially on old plants producing sectors in the old wood. The reason is that the roots do not thicken and produce an annual ring during years when their vascular bundles do not connect them directly with above-ground stems.

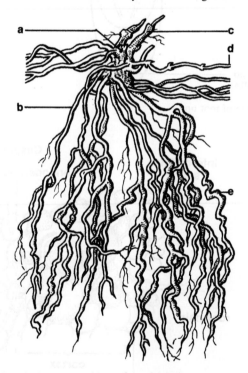

Fig. 9. Displacement of vertical roots: a – old wood, b – vertical roots, c – new wood, d – suckers, e – root tubers.

According to their spatial position the skeletal roots are classified as vertical and horizontal. Both types are injured by pests and diseases and horizontal roots may also be damaged by mechanical cultivation of the soil. The oldest roots become spontaneously atrophied, a process which usually starts in the oldest tissues within the basal part of the root and proceeds to the cambium. After the partial decay of the cambium, the root splits into two or more parts, some of which atrophy. The main functions of the root are then limited, and once the root has decayed they completely disappear.

V e r t i c a l (t a p) r o o t s include all skeletal roots and their branches growing vertically or nearly so. The vertical roots are the first roots produced by the new hop plants, and one tap root grows from each germinating seed. In vegetative propagation a large number of tap roots is usually produced at the base of the transplanted cutting. These are designated as rootstock tap roots because they grow from the old wood of rootstock organs.

In addition, tap roots can also grow from the horizontal roots. These are designated as *anchor roots*, and up to 30 rootstocks and anchor tap roots can be found, on a well-grown vegetatively propagated hop plant. The largest number of rootstock tap roots occurs on young plants aged from one to three years. These anchor roots are produced stepwise, starting when a branch growing from horizontal roots becomes thickened vertically. The anchor roots are not produced solely from the need to anchor the plant but, originally, so as to take in water from the deeper soil layers, as the activity of old tap roots becomes reduced (RYBÁČEK, 1978).

H o r i z o n t a l b r a n c h r o o t s may be produced by all organs of the rootstock, by old as well as new wood by suckers and by vertical roots. Such roots, growing from a rootstock organ are designated as *rootstock roots* and those originating from vertical roots are called *floor roots*. All lateral roots have a high potential for branching. According to PROČAJEV (1955) they reach sevenfold branching so that they ramify densely in the soil layers. Under

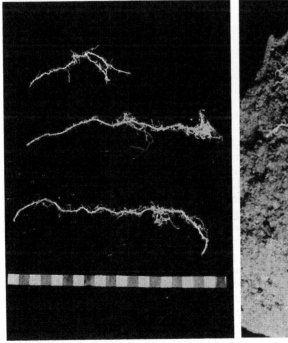

Fig. 10. Horizontal roots.

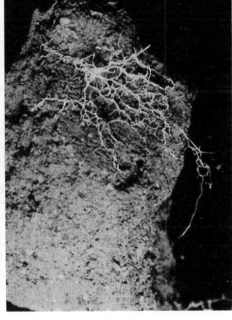

Fig. 11. Branching of horizontal roots.

39

favourable conditions damaged lateral roots are quickly replaced. The profuse branching and intensive activity of horizontal roots led certain scholars to conclude that their activity alone affects the yield rate. Thus, MOHL (1924) asserted that nutrients from the soil are provided exclusively by these roots (one year old, and therefore called summer roots) growing from new wood. This view can be contradicted by the fact that new horizontal roots can become vertical anchor roots (RYBÁČEK, 1978) when the function of old vertical roots is lost or temporarily checked.

The rootstock roots and the horizontal floor roots can regenerate very quickly and therefore this part of the root system is usually younger than the vertical roots.

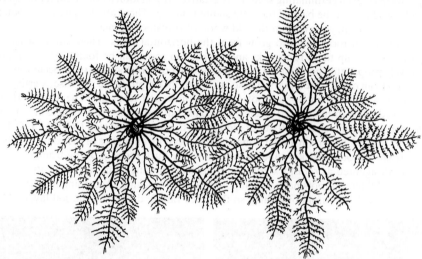

Fig. 12. Displacement of horizontal roots.

Root tubers

Root tubers are secondary thickened roots which serve as the main storage organ. They occur on vertical root branches about 30–40 cm below the surface of the soil and they are usually bottle-shaped. Their histological structure is characterized by hypertrophied cortical tissues. A cross section clearly shows the developed parenchymatous tissue of the secondary cortex where the reserve substances, mostly starch and sugars are stored. The hypertrophic parenchymatous tissue makes the cortex quite elastic, enabling it to adapt to the soil environment, and this accounts for the many variations in shape of root tubers. These deformations seem least to involve the ends of the tuber, resulting in its bottle shape. These deformations often affect the cross section causing it to be irregular and often oval with an eccentrically placed central cylinder.

After the reserves, especially starch and sugar have been exhausted, the parenchymatous tissue and the epidermis become atrophied, thus ending the main function of the root tuber in a given year. New root tubers are usually produced in new positions on younger vertical root branches. If the tubers are formed on all vertical roots their diameter is smaller than those produced on young roots . Because root tubers usually develop on young vertical roots and have hypertrophic cortical tissues, they are, biologically, very young root organs. The extinction of the storage tissue and the epidermis of exhausted tubers makes their replacement more rapid than that of vertical and horizontal roots .

Fig. 13. Root tuber.

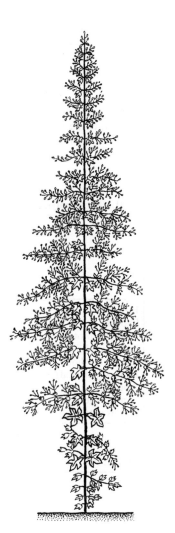

Fig. 14. Pattern of above-
-ground part of the hop plant.

STRUCTURE AND FUNCTION OF THE ABOVE-GROUND PARTS OF THE HOP PLANT

Two separate organ systems – vegetative and generative – arise successively in the above-
-ground part of the hop plant. These can be distinguished by the morphological structure of
their organs and also by their main functions.

The s y s t e m o f v e g e t a t i v e o r g a n s , consisting of stems and leaves, has as its main
function the production of organic substances by photosynthesis and connected procedures.
These products are temporarily accumulated in reserve tissues, namely the stems.

The main function of the g e n e r a t i v e o r g a n s is the production of seeds in the hop
fruit. But in the case of the cultivated hop this function is suppressed by means of an efficient
isolation of cultivated female stocks from any wild male plants growing in the locality, thus
preventing fertilization and fructification.

The two above-ground systems not only have different structures and functions but there is a close connection between the generative system and almost fully-developed vegetative system. The generative organs arise from terminal buds of the vegetative system, which stops any further growth of the stem and its branches in the fertile parts of the plant.

The main differences in chemical composition between the two systems, at the time of technical maturity, are shown in Tables 9, 10, 12 and 13. These show that the generative organs (cones) have a much greater content of ether extract, more nitrogenous substances and less ash, crude fibre and extracted nitrogen-free substances. Thus the biological age of the cones, at the time of their technical maturity, is lower than that of the above-ground vegetative system.

TABLE 9a **Analysis of above-ground vegetative organs of five-year old hop plants**

Plant part	Constituents (percentage of dry matter)					
	ash	nitroge-nous substances	ether extract	crude fibre	nitrogen free extractive substances	total organic substances
Bines	6.14	5.30	1.35	38.08	49.13	93.86
Shoots	6.89	5.63	1.37	39.90	46.21	93.11
Stems (subtotal)	6.36	5.40	1.36	38.62	48.26	93.64
Leaves on bines	22.04	13.34	4.75	11.92	47.95	77.96
Leaves on shoots	20.29	15.05	4.37	11.62	48.65	79.71
Leaves (subtotal)	20.99	14.37	4.52	11.74	48.38	79.01
Above-ground system of vegetative organs in total	12.82	9.36	2.76	26.75	48.31	87.18

TABLE 9b **Analysis of above-ground vegetative organs of five-year old hop plants**

Plant part	Constituents (percentage of dry matter)					
	starch	reducing sugars	other nitrogen free extractive substances	nitrogen free extractive substances	glucose from starch	total monosac-charides
Bines	1.83	7.20	40.10	49.13	2.03	9.23
Shoots	2.20	2.30	41.73	46.21	2.44	4.74
Stems (subtotal)	1.94	5.74	40.58	48.26	2.15	7.89
Leaves on bines	0.08	1.37	46.50	47.95	0.09	1.46
Leaves on shoots	0.02	2.50	46.13	48.65	0.02	2.50
Leaves (subtotal)	0.04	2.05	46.29	48.38	0.05	2.10
Above-ground system of vegetative organs in total	1.10	4.11	43.10	48.31	1.22	5.33

TABLE 10a **Dry matter content and major elements in dry matter of above-ground vegetative organs of five-year old hop plants**

Plant part	Dry matter				Dry matter content			
	in total (g)	percentage	ratio	C (%)	N (%)	C/N (C:N)	10 Ca/N (Ca:N)
Bines	132.00	24.00	39.22	46.930	0.848	55.342	19.835
Shoots	55.90	31.05	16.61	46.555	0.901	51.670	20.355
Stems (subtotal)	187.90	25.79	55.82	46.818	0.864	54.250	19.990
Leaves on bines	59.50	23.80	17.68	38.980	2.134	18.266	20.281
Leaves on shoots	89.20	27.87	26.50	39.855	2.408	16.551	16.445
Leaves (subtotal)	148.70	26.57	44.18	39.505	2.298	17.237	17.980
Above-ground system of vegetative organs in total	336.60		100.00	43.587	1.498	37.899	19.102

TABLE 10b **Dry matter content and major elements in dry matter of above-ground vegetative organs of five-year old hop plants**

Plant part	Dry matter content						
	Cl (%)	S (%)	F (%)	Ca (%)	K (%)	Mn (%)	Fe (%)
Bines	0.446	0.088	0.274	1.682	1.536	0.207	0.043
Shoots	0.243	0.102	0.253	1.834	1.144	0.203	0.067
Stems (subtotal)	0.386	0.092	0.268	1.727	1.419	0.206	0.050
Leaves on bines	0.530	0.300	0.296	4.328	1.565	0.372	0.044
Leaves on shoots	0.589	0.328	0.312	3.960	1.504	0.385	0.085
Leaves (subtotal)	0.565	0.317	0.306	4.107	1.528	0.380	0.069
Above-ground system of vegetative organs in total	0.465	0.191	0.285	2.778	1.467	0.283	0.058

System of above-ground vegetative organs

The above-ground vegetative organs include the *buds,* the *stem* and the *leaves.* The stem, produced by the successive development of new tissues from vegetative apex, constitutes the skeleton of the whole vegetative system of the hop plant. The different functions of particular vegetative organs affect their morphology and also their *chemical composition,* as shown in Tables 9 and 10. These indicate essential differences between leaves and stems, less important differences between bines and shoots, and similarly smaller differences between leaves on them.

Above-ground buds

The *terminal bud* of the stem arises underground on an organ of the rootstock. During growth its apex is enveloped in a large number of bracts which become green once apex rises above the ground. *Leaf buds* occur in the axils of the bracts. Individual tissues are subsequently differentiated at the apex and serve as a base for all of the above-ground organs, including new lateral buds. *Dormant lateral buds* occur in the axils of leaves and some of them proceed to further growth. Each leaf is usually accompanied by three dormant buds. Any suppression or interruption of apical dominance activates the central bud which will then suppress its neighbours. Lateral branches, as well as their secondary elements, grow from these central buds.

The generative organs grow from the apical buds of fertile stems and their branches. They are also produced from the two lateral buds in the leaf axils around the fertile shoot at the apex of the stem.

Vegetative buds become modified into generative organs or disappear due to atrophy. Active as well as dormant buds on sterile stems and shoots decay successively from the apex downwards. RYBÁČEK (1964) discovered that the extinction of buds precedes the extinction of other above-ground organs. This phenomenon was used to establish the phenological phases of above-ground parts taken from the hop plant after the cones had been picked.

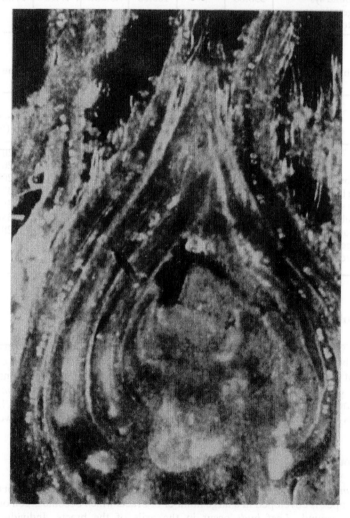

Fig. 15. Grown apex of hop in detail.

Stems

Stems form the basis of the above-ground system. In the hop plant the stem includes the *main bine* and its *lateral branches (shoots)*. The main stems arise as the first above-ground organs in spring and are the last to decay down to ground level in the autumn. If the stems can twine around a suitable support, they can grow, during the growing season, up to a length of 8 or 9 meters, while at the same time becoming thicker. According to MOHL (1924) only the 3 or 4 apical internodes of h o p s t e m s can elongate. The lower, older internodes cease to increase in length, but continue to thicken and when this ceases then only those changes occur which are involved in tissue ageing. The formation of all tissues does not end during the growth of the upper internodes and nodes. Only the nodes are entire in the older parts of the stem because the cortex of the internodes is torn during thickening, thus causing a pith cavity. The internal structure of a well-grown hop stem can thus be observed only in cross and longitudinal sections through the center of a node. The whole surface of the stem is covered by bark (epidermis) built from a single layer of tangentially elongated cells the membranes of which, especially on the outer side, are very thick. The cortical tissue is attached to the inner side of epidermis and the parenchyma with some remnants of heartwood to the inner cavity. The cortical and lignified part of the stem is divided by a thin layer of dividing tissue, the cambium. The xylem parenchyma forms a wide, most regular, hexagon. Because the cells of xylem parenchyma soon decay, this hexagon affects the outer shape of internodes throughout their life. Along the edges of the hexagon the vascular bundles occur in groups of 3 or 4 so that in the whole stem they total 18 to 24. Phloem occurs on the outer side and xylem on the inner. Thick-walled tubular vessels occur in the phloem and these distribute mobile substances from the leaves. The xylem contains many vessels (tracheae and tracheids) which distribute solutions from the roots.

The relatively thick cortical part of the stem consists of many layers of parenchyma. These layers, constituted from thin-walled cells, are stiffened by a compact layer of sclerenchyma forming a hexagon inside the bark. This sclerenchymatous hexagon has a greater diameter than the xylem, with which it participates in the hexagonal structure of the stem. The corners of the hexagon sometimes protrude. Hairs, consisting mostly of silicates, grow on all six edges and as these harden they become one or two sided hooks. The stems attach themselves by these hard hooks, even to relatively smooth supports. The sclerenchyma in the bark increases

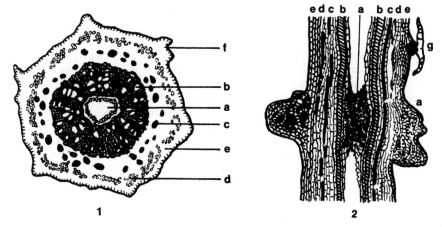

Fig. 16. Cross (1) and longitudinal (2) section of a stem: a – heartwood, b – lignified cylinder, c – vascular bundles of phloem, d – sclerenchyma fibres, e – cortex, f – epidermis, g – hook.

the elasticity and strength of stems. Their torsional end bending elasticity is further improved by collenchyma under the epidermis at the corners. The cells in the outer layers of the parenchymatous tissue below the epidermis contain chlorophyll and under favourable conditions they can perform photosynthesis. Cells in inner layer, being without chlorophyll, participate only in the distribution of substances.

The structure of the lateral shoots is broadly similar to that of the stem, but in some parts it is reduced. The shoots arise from buds in the axils of leaves when the stem has reached a length of 100 to 200 cm. L a t e r a l s h o o t s also extinct earlier than the stem. According to RYBÁČEK (1963) the chemical composition of the stem was little different from that of lateral shoots (in five-year old plants of 'Osvald's clone No. 72', picking date August 28, 1963).

The greater content of dry matter, ash and crude fibre and less nitrogen-free substances implies that the biochemical changes in the shoots reached a higher or later stage than that of the stems. The analysis of stems and their respective lateral shoots or branches should also show differences in chemical composition. These differences were established by RYBÁČEK (1963) for starch and reducing sugars, as shown in Table 11.

TABLE 11   **Content of reducing sugars and starch in dry matter from the stem of hop plant (per cent)**

Upper part of the stem						Stem of the whole plant	
Height 8-50 cm		Height 50-300 cm		Height 300-720 cm			
Reducing sugars[x]	starch	reducing sugars	starch	reducing sugars	starch	reducing sugars	starch
4.88	0.70	4.34	1.14	6.64	2.35	5.74	1.93

[x] Sugars reducing Fehling's solution.

Leaves

Leaves grow from the nodes of main bine lateral branches in pairs normally opposite each other. According to their position the *leaves of stem (bine)* and *of shoots* are different. The leaves on the bine develop earlier, are usually bigger and have a rougher structure. Both types are petiolate, but the petioles of bine leaves are thicker. The blades of young leaf are much folded, but well-grown leaves are gently corrugated. The margin of leaves is roughly serrated or palmately lobed. Young leaves are cordate, though this shape is not retained by well-grown leaves. Well-grown main bine leaves have five or seven lobes, whereas leaves on lateral branches are compact or with three or five lobes. The venation of leaves is palmate. The main veins protrude from the undersurface of the leaves. Silicate hooks identical with those on the petioles and bines occur on the underside. The upper surface is a deeper green than that of the reverse. Light-coloured glands containing resins and essential oils are found on the reverse.

MOHL (1924) drew attention to the content of resins and essential oils in internal lenticular glands, found as nodes within the venation of leaves. DARK and TATCHELL (1955) found a direct relationship between the number of resinous glands in the leaves and the resin content of hop cones. This relationship can be used in breeding for a higher resin content in hop cones (see Fig. 17).

The anatomical structure of hop leaves is analogous to that of shade loving plants. They have a thin layer of mesophyll with a palisade parenchyma of only one stratum consisting of dilated cells with a thickness to length ratio of 1:6. The upper epidermis consists of large rectangular cells covered with a thin cuticle. The palisade layer contains the cells transferring the assimilates and below it, the layers of spongy parenchyma have globular cells and large intercellular spaces. The lower epidermis is made up of small cells covered with thin cuticle penetrated in places by cellular pores (stomata). There are about 400 stomata per mm^2 on the underside of leaves providing for very intensive metabolism.

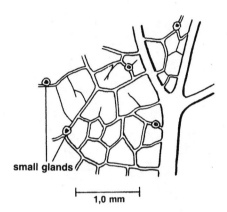

small glands

1,0 mm

Fig. 17. Lupulinic glands in hop leaves.

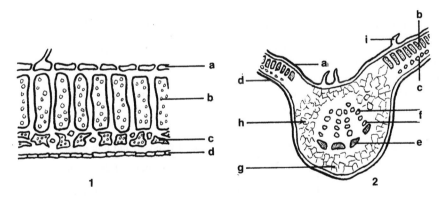

Fig. 18. Structure of leaf (1) and central vein (2): a – upper epidermis, b – palisade parenchyma, c – spongy parenchyma, d – lower epidermis, e – phloem, f – veins, g – basic parenchyma, h – xylem, i – hook.

The differences in the major components of the main bine and lateral branch leaves are shown in Tables 9 and 10.

The early main bine leaves and later lateral branch leaves show differences, at the hop picking period, not only in their calendar age but also in their biological age. This also affects their chemical composition (lateral branch leaves contain more nitrogenous substances and extracted nitrogen-free substances, less ash and ether extractables).

47

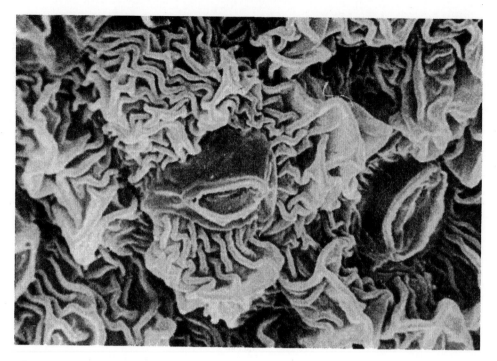

Fig. 19. Stomata of hop leaves.

The generative organs

The apical of the main bines and the buds of the fertile lateral branches produce *inflorescences* which on female plants become the *syncarps*. The fertilized flowers produce the *fruits* which are *one-seed achenes* known in hop-growing terminology as *"stones"*. Similarly the inflorescence is called "envelope" and the syncarp the "cone". The inflorescence becomes the syncarp when the stigma is no longer able to receive pollen. This change occurs even when the flowers are not fertilized, so that cones are produced by female plants without fertilization, i.e. parthenocarpically. *The differences in chemical composition between inflorescences, cones at different stages of maturity, and hop fruits are shown in* Tables 12 and 13.

TABLE 12a   **Analysis of generative organs and buds of five-year old hop plants**

Generative organs	Date	Constituents (percentage of dry matter)					
		ash	nitro-genous substances	ether extract	crude fibre	nitrogen free extractive substances	total organic sub-stances
Inflorescence	28/7	12.10	23.93	4.35	11.78	47.84	87.90
Cones	8/8	12.13	18.9'	6.21	11.04	51.71	87.87
Cones	28/8	8.60	20.63	18.96	15.72	36.04	91.40
Cones	19/9	9.01	11.04	18.34	24.14	37.47	90.99
Fruits	19/9	6.10	38.82	31.21	9.73	14.14	93.90
Buds	19/9	9.53	40.25	–	6.25	43.97	90.47

TABLE 12b Analysis of generative organs and buds of five-year old hop plants

Generative organs	Date	Constituents (percentage of dry matter)					
		starch	reducing sugars	other nitrogen free extractive substances	total nitrogen free extractive substances	glucose from starch	total mono-saccharides
Inflorescence	28/7	–	2.50	45.34	47.84	–	2.50
Cones	8/8	0.05	5.90	45.76	51.71	0.055	5.95
Cones	28/8	0.03	3.00	33.01	36.04	0.033	3.03
Cones	19/9	–	2.25	35.22	37.47	–	2.25
Fruits	19/9				14.14		
Buds	19/9	–	2.02	41.95	43.97	–	2.02

TABLE 13a Major elements in dry matter of generative organs and buds of five-year old hop plants

Generative organs	Date	Dry matter content					
		C (%)	N (%)	C/N (C:N)	10 Ca/N (Ca:N)	Cl (%)	S (%)
Inflorescence	25/7	43.950	3.829	11.478	3.965	0.099	0.334
Cones	8/8	43.94	3.026	14.521	4.465	0.220	0.324
Cones	28/8	45.70	3.310	13.807	3.891	0.230	0.344
Cones	19/9	45.50	1.766	25.764	9.541	0.201	0.301
Fruits	19/9	46.950	6.211	7.559	0.343	not investigated	
Buds	19/9	44.960	6.44	9.386	0.533	0.077	0.327

TABLE 13b Major elements in dry matter of generative organs and buds of five-year old hop plants

Generative organs	Date	Dry matter content				
		P (%)	Ca (%)	K (%)	Mg (%)	Fe (%)
Inflorescence	25/7	0.852	1.415	3.095	0.274	0.203
Cones	8/8	0.829	1.351	3.156	0.272	0.176
Cones	28/8	0.564	1.288	3.284	0.247	0.068
Cones	19/9	0.483	1.685	2.412	0.242	0.066
Fruits	19/9	not investigated	0.213	not investigated		
Buds	19/9	0.818	0.343	4.659	0.280	0.235

Inflorescence

The climbing hop is dioecious i.e. the individual plant is either male or female. Male and female inflorescence on one branch, as in monoecious plants is very rarely seen.

The male inflorescence is a richly branched panicle with small individual flowers on short stalks. At blossoming they reach a diameter of 5–6 mm. They have 5 petals with adherent stamens and anthers bearing fine yellow pollen. The petals are also provided with lupuline glands but their number is smaller than on the female inflorescence.

The female inflorescence consists of 20 to 60 flowers on a cranked axis. Each crank usually bears two pairs of flowers (exceptionally only one) and each pair of flowers is protected by a single covering leaf originating from a bract. The authentic leaf, in the

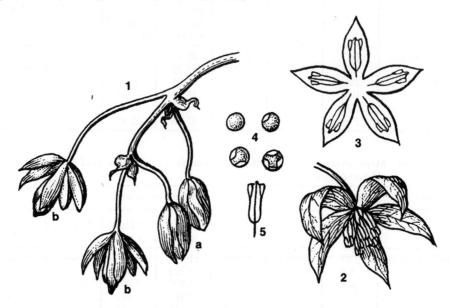

Fig. 20. Male inflorescence (1) with buds (a) and flowers (b), open flower (2) and flower diagram (3), pollen grains (4), anther (5).

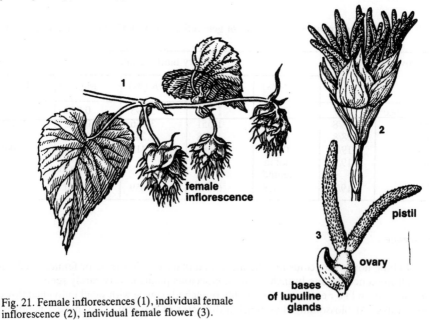

Fig. 21. Female inflorescences (1), individual female inflorescence (2), individual female flower (3).

inflorescence, is only rudimentary and is very small and bunched. Individual flowers are situated in the axils of the bracts.

Each individual flower has a minute, green perianth which closely encompasses the unilocular bicarpelate ovary with one ovule. The ovary bears two filamentous stigmas, without styles. Both stigmas are fixed only near the micropyle (seed aperture), elsewhere they are free and on their surface are covered by long papillae which catch the pollen floating in the air. The stigmas are whitish and as they atrophy they very quickly become brown and lose the ability to catch the pollen. With this loss their function ends, as does that of the flower. The whole inflorescence then changes to syncarpy.

The inflorescences grow from *flower-buds*. The time, when the stigma appears over the edge of the covering leaf is taken as the beginning of inflorescence. The development of the inflorescence is in four stages (see Fig. 47). Cultivated hop plants blossom in waves. If a rainy period follows dry time during flowering, then those plants which have already blossomed begin to blossom again, i.e. to produce a second inflorescence. The number of flowers in this case is usually lower. The stocks of cultivated hop have certain naturally-occurring abnormalities in their inflorescences. Well-grown plants which do not blossom are designated as sterile, those which start to flower early in the season are know as "early blossoming", and those with a prolonged blossoming period are long blossoming. Plants with these abnormalities are removed from stocks and replaced with new normal and healthy individuals.

Fig. 22. Well-grown inflorescence.

Syncarpy – hop cone

Hop cone (botanically, hop strobile) is, morphologically, an inflorescence which originated from a catkin. Developmentally it is a shortened and modified branch. As well as the stem (stalk) which is modified into the *rachis* and the *cranked axis,* the essential part of the cone, according to KAVINA (1925) and OSVALD (1940), consists of *bracteoles* and *bracts* modified into *covering bractlets* and *true (involucral) bracts.* Special features typical of hop cones are the outer *lupulinic glands, the oneseeded fruits (the achenes or so-called "stones")* and *the unfertilized fruits, without seeds (so-called "small stones").* As with the inflorescence four stages are recognized on growing cones.

The relative amounts of individual parts of the cone in dry well-grown seedless cones differ according to hop variety and also in different growing conditions as is shown in Table 14.

Commercially the most important component of the hop cone is its l u p u l i n e . This is the designation given to the multicellular l u p u l i n i c g l a n d s , which are produced by epidermal cells. They are cup-shaped or globular and contain resins and essential oils which

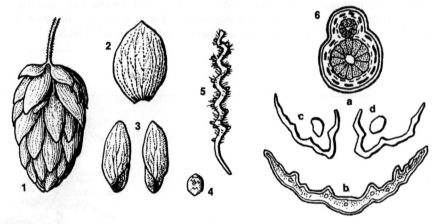

Fig. 23. Cone and its parts: 1 – cone, 2 – covering bract, 3 – true (involucral) bracts, 4 – achene, 5 – cranked axis, 6 – cross section of a cone |a – cranked axis, b – covering bract, c – true (involucral) bracts, d – achenes|.

TABLE 14 **Percentage of components in dry matter of hop cones**

Component	Percentage of total				
	Wildner (1938)			Bulgakov (1954)	Rybáček (1963)
	min.	max.	average		
Rachis	5.20	5.90	5.60	5.98	4.40
Axis	6.10	7.30	6.70	6.36	6.68
Bracteoles and bracts	66.90	69.40	68.20	66.85	69.31
Lupuline glands	19.20	19.80	19.50	19.81	19.61
Total			100.00	100.00	100.00

Note: The percentage of rachis in the last column is lower because the analysis included cones picked by machine (RYBÁČEK, 1963).

Fig. 24. Mature hop cones.

Fig. 25. Cranked axis and bracts: 1 – fine and regular cranked axis of cultivated varieties, 2 – axis of rougher varieties, 3 – axis of wild hops, 4 – arrangement of bracts on crank of an axis, a – covering bractlet, b – true (involucral) bracts|.

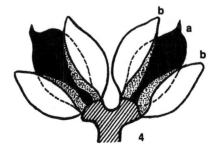

53

give the golden-yellow colour to lupuline. As well as the external lupuline glands there are elongated lenticular filled with lupulinic material on the inside of the bracts.

These inner chambers add 15–20 per cent to the content of lupuline. Seedless cones produce more lupuline and its quality is higher. The determination of lupuline is very difficult and therefore indirect methods of analysis are used. With a simple mechanical analysis the content of lupuline and its quality depend on the structure and composition of the cone. In particular, the composition and properties of the axis, as well as the number and properties of bracts are important. Chemical analyses, however, are more precise, because all of the resin contained

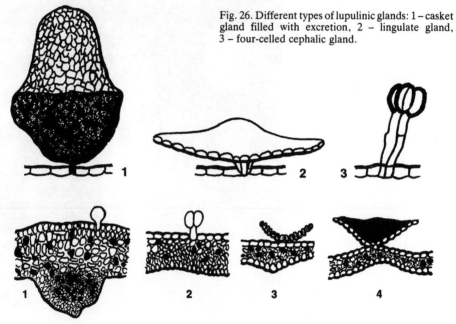

Fig. 26. Different types of lupulinic glands: 1 – casket gland filled with excretion, 2 – lingulate gland, 3 – four-celled cephalic gland.

Fig. 27. Development of lupulinic gland (1–4) on bracts of the hop cone.

TABLE 15 **Analysis of hop cones**

Constituent	Content (percentage of dry matter)				
	Mohl (1924)	Hrach (1928)	Rybáček (1963)	Burgess (1964)	Zattler and Maier (1966)
Ash	8.57	8.77	9.01	8.89	8.5
Total nitrogenous substances	20.00	22.80	21.04	16.87	20.0
Ether extract	21.37	21.05	18.34	16.66	·18.3
Crude fibre	15.20	16.37	14.14	–	15.0
Nitrogen-free extractive sub-stances	34.86	31.01	37.47	–	38.2
Fibre and nitrogen-free extractive substances	50.06	47.38	51.51	57.58	53.2

in the cone is involved, most of it (up to 80 per cent) in lupuline and considerably less (only 20 per cent) in the other parts (rachis, cranked axis and bracts).

Complex chemical analyses in Czechoslovakia as well as abroad indicate various changes in the content of major groups of substances as shown in Table 15. In particular the decrease in ethereal oils (total resins) and crude fibre is remarkable. During the same period the content of extracted nitrogen-free substances has increased. Such results imply that less ripe cones were picked in Czechoslovakia during the 1960s.

Chemical substances and their production in hop cones

The chemical substances in hop cones are classified, with regard to their brewing properties, into two main groups:
a) efficient substances of primary importance for beer production;
b) accompanying substances of secondary importance.

The efficient substances include resins (bitter substances), essential oils (hop oil) and polyphenol substances (tannin). These three kinds of substances are better known than the accompanying substances, which are secondary only with regard to beer production. These are important in hop production and commerce, because their quantity and composition depends on the origin of the hop, the ecological conditions and agrotechnical measures taken for breeding.

The chemical and biochemical analyses of hop cones pay most attention to the resins, tannin and oils, so that the levels of other substances tend to be ignored. Therefore the available literature gives only sparse information about them even though some of them are important from breeding point of view.

Hop resins are a complicated complex of nitrogen-free substances. As discussed in Chapter 1 (Utilization of hop products) they are classified into soft and hard resins. The soft resins form a group of α-bitter acids and the so-called β-fraction. The relative amount of these fractions in the various parts of the hop cone is shown in Table 16, as determined by BULGAKOV (1954). From the breeding point of view, knowledge concerning the location of hop resins in plants, their changes and modifications during plant growth, after-harvest and during storage is very important.

TABLE 16 **Chemical substances in the individual components of hop cones**

Component	Content (percentage of dry matter)				
	total resins	soft resins	α-bit-ter acids	β-fractions	hard resins
Lupuline glands	79.55	77.40	34.61	42.79	2.15
Bracts	7.44	7.00	3.10	4.00	0.34
Cranked axis ("strig")	3.41	3.28	1.44	1.84	0.13
Rachis	1.77	1.71	0.72	0.99	0.06

According to recent research, the production and transformation of hop resins has been shown to occur in four steps: At first phloroglucine is produced, followed by β-bitter acids and α-bitter acids, and these are transformed into β-hard resins.

a) Formation of the phloroglucine system

Investigations into the biosynthesis of the phloroglucine nucleus, have used radioactive carbon (C^{14}). These have shown that the nucleus is formed from three molecules of acetic acid, and after the elimination of two molecules of water the β, γ-diketo-caproic acid is produced. This acid combines with the acyl group of α-keto-acids produced from the transamination of certain amino-acids, including valine, leucine and isoleucine. Thus, valine produces α-keto-isovaleric acid and leucine gives α-keto-caproic acid. Their acyls combine with particular β homologues and α-bitter acids. Among those already isolated are the following: isovaleryl for lupulone and humulone, isobutyryl for the co-components (colupulone and cohumulone), methyl-ethyl-acetyl for the ad-components, isocapronyl for the pre-components and n-propyl for the post-components: these are for the five know β homologues and α-bitter acids (the theorical number is seven). A synoptic diagram, according to STOCKER (1963), is shown in Fig. 28.

b) Formation of β-bitter acids

Prenylation of the phloroglucine nucleus produces β-bitter acids. Their homologues are distinguished by their bonded acyl group. At present, five homologues have been identified, namely: lupulone, colupulone, adlupulone, prelupulone and postlupulone. Their ratio depends on the properties of the hop variety and of the prevailing ecological conditions. According to VANČURA (1966) technically mature hop cones in the Žatec hop breeding region contain approximately 60 per cent of lupulone, 30 per cent of colupulone and 10 per cent of adlupulone. The content of prelupulone and postlupulone is insignificant.

c) Formation of α-bitter acids

The β-bitter acids are transformed by solar radiation (photolysis), first to 4-desoxy-β-acids, then to α-bitter acids as shown in Fig. 28. This hypothesis by STOCKER (1963) has been verified by the chemical analysis of hop cones as they ripen. During the initial phases the content of β-bitter acids in cones is relatively high whilst the content of α-bitter acids is low. As ripening progresses, and especially in overripening, the content of α-bitter acids increases. The fact that cones in the un-shaded top parts of the plants contain more α-bitter acids than those produced earlier in the shaded middle and lower parts shows the profound effect of solar radiation. Thus, the content of α-bitter acids depends not only on the length of the ripening period but also on the intensity of illumination. As is the case with β-bitter acids, five homologues of α-bitter acids have been identified, namely humulone, cohumulone, adhumulone, prehumulone and posthumulone.

Technically ripe cones of all authorized Czechoslovak varieties have the following average composition: 70 per cent of humulone, 20 per cent of cohumulone; approximately 8 per cent of adhumulone and about 2 per cent each of prehumulone and posthumulone. The content of humulone and cohumulone in technically ripe cones is used as an important, genetically fixed, varietal character.

RICHBY (1958) classified the most important hop varieties into four groups according to the percentage of cohumulone in their total α-bitter acids:

1. cones with 20 per cent cohumulone – all Czechoslovak varieties and the West German varieties 'Tetnang', 'Spalt' and 'Hallertauer';

2. cones with 30 per cent cohumulones – the English varieties in the groups of 'Fuggles', 'Goldings' and 'Northern brewer' and the American varieties 'Oregon Fuggles', 'Oregon Seedless' and 'Canadian Pride';

3. cones with 40 per cent cohumulone – varieties 'Brewers Gold', 'Early Promise', 'Oregon Cluster' and 'Washington Seedless';

4. cones with more than 40 per cent cohumulone – varieties 'Bullion Hop', 'Yakima Seedless' and 'Californian Seedless'.

Highest content of cohumulone (65 per cent) was found in a hop from New Mexico (*Humulus lupulus* L., subspecies *neomexicanus*, Nels. et Cockerell).

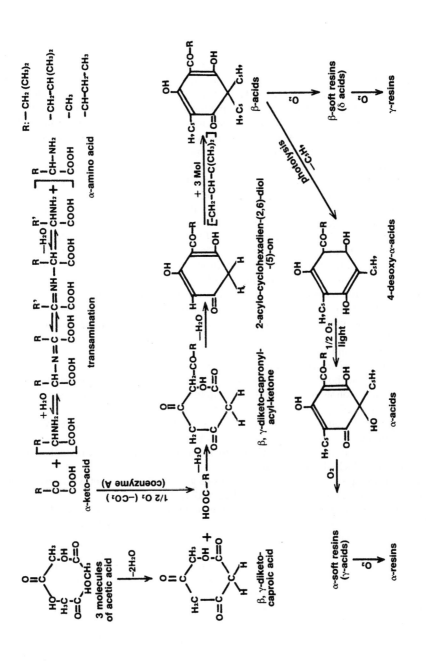

Fig. 28. Biosynthesis of bitter substances according to STOCKER (1963).

d) Formation of other soft and hard resins

During the later stages of transformation the α- and β-acids become oxidized and other soft resins (γ-bitter acids = humulinones, δ-bitter acids = hulupones) result from their condensation and polymerization. Finally, the hard resins are produced (δ-hard resins and γ-hard resins).

It has been found that the oxidation of α-bitter acids is earlier and quicker than that of β-bitter acids. This may be one of the reasons why cones of Czechoslovak hops become old very slowly, because they have more β-bitter acids than α-bitter acids. GOODCOP (1954) found the best resistance to ageing in the variety 'Žatecký' (hops from the Žatec region). The whole process of hops growing old, with its accompanying transformations, is not yet completely understood. According to DYR and HAUZAR (1962), the ratio of particular homologues of α- and β-acids does not change either during the storage of hop cones nor during the brewing of beer. They enter beer in practically the same ratio as they are present in the hop cone.

Some humulone preparations have been synthesized in the laboratory but they cannot replace the extensive complex of natural substances contained in hop cones. Therefore the European convention of brewers recommended in 1967 that only preparations produced from the natural hop, and containing only natural substances should be used in the brewing industry, and not artificially synthesized products.

H o p   e s s e n t i a l   o i l contains volatile substances, the composition and properties of which account for the scent of hop cones. This provides one of the many important indicators for the estimation of the quality and origin of hop cones. More than 200 different substances have been identified in hop essential oil. They are divided into hydrocarbon and oxygen fractions. This latter contains oxidized hydrocarbons and is known as the oxy-fraction.

The hydrocarbon (terpene) fraction includes:
1. aliphatic hydrocarbons (e.g. n-pentane, n-octane, isoprene);
2. monoterpenes (e.g. myrcene, β- and γ-pinene, ocimene and limonene);
3. sesquiterpenes (e.g. humulene, farnesene, β- and γ-caryophylenes, δ- and γ-hadinene, β- and γ-selinene);
4. triterpenes and others (e.g. aqualene).

The oxygen (oxidated) fraction comprises:
1. aliphatic alcohols (e.g. methyl alcohol, pentyl alcohol and isopentyl alcohol);
2. terpenic alcohols (e.g. humulenole, linelole, geraniole, nerolidole and terpinole);
3. other alcohols (e.g. phenyl-ethyl alcohol);
4. aliphatic ketones (e.g. all homologous methylketones from dimethyl-ketone up to hexadekanone with branched as well as non-branched chains);
5. aliphatic aldehydes (e.g. propylaldehyde and further homological aldehydes up to C_{11});
6. terpenic aldehydes (e.g. citral);
7. esters of alphatic and terpenic alcohols (e.g. methyl esters, methylthioesters, butyrates, acetates, propionates, capronates and heptanoates).

The ratio between the hydrocarbon and the oxygen fraction differs according to variety, as shown in Table 17, where the results of analyses made by SILBERREISEN and KRÜGER (1967) are presented.

The variety 'Žatecký' has the least hydrocarbon fraction (33 per cent) and greatest oxygen fraction (67 per cent). The most important of the hydrocarbon fraction are the monoterpene, myrcene and the sesquiterpenes, humulene, caryphylene and farnesene. Their content in European hops, where they make up to 90 per cent of the hydrocarbon fraction, is shown in Table 18. Certain English, American and Australian varieties have less of these terpenes because in these hops, selinenes are prevalent, ranging up to 40 per cent in the hydrocarbon fraction. According to recent work, these hop cones with the lower content of the monoterpene, myrcene and a higher content of sesquiterpenes like humulene, posthumulene,

58

TABLE 17 **Content of α-bitter acids, and the total essential oil with its main fractions in hop cones of named varieties**

Variety of hop	Water content (per cent)	Content of α-bitter acids (per cent)	Estimation of flavour (score)	Content of essential oil in dry matter (per cent)	Percentage in total essential oil	
					hydrocarbon fraction	oxygen fraction
'Hallertauer' (West-Germany)	12.2	5.6	28	1.83	57.94	42.06
'Tettnang' (West-Germany)	9.0	6.5	27	1.45	48.45	51.55
'Spalt' (West-Germany)	9.8	4.8	20	1.47	50.31	49.69
'Jura'	10.6	5.9	27	1.75	50.92	49.08
'Hersbruck'	7.8	5.3	16	1.50	49.15	50.85
'Žatecký'	9.0	5.5	24	1.56	34.88	65.12
'Steirer'	12.2	6.9	23	1.74	49.72	50.28
'Vert d'Alsace'	8.3	5.2	9	1.66	50.58	49.42
'Northern Brewer'	9.0	10.1	17	2.43	63.43	26.57

TABLE 18 **Major components of the hydrocarbon fraction in essential oils – different European hops**

Variety of hop	Content of components (mg per 100 g dry matter)			
	myrcene	caryophylene	farnesene	humulene
'Žatecký'	140	36	114	160
'Tettnang' (West-Germany)	223	43	173	185
'Vert d'Alsace'	246	89	99	267
'Spalt' (West-Germany)	214	99	86	322
'Steirer'	324	125	68	200
'Nersbruck'	374	123	4	253
'Jura'	262	145	1	499
'Hallertauer' (West-Germany)	246	134	1	540
'Northern Brewer'	59P	134	1	600

β- and δ-caryophylene and farnesene are suitable for the production of quality beers. Such qualities are found in the variety 'Žatecký', which also has more oxidized hydrocarbons.

The distribution of the essential oil in hop cones is similar to that of the resins. It is found in greatest quantity in lupuline. As is clear from Fig. 29, the terpene fraction of the oil is produced from the same basic substance, i.e. from acetic acid. During the growth of cones, and during their post-harvest processing and storage, the two groups of active substances, resins and oils, influence each other. Analyses have shown that the contents of myrcene and cohumulene in hop cones of particular varieties correspond and that they develop together in the complex of α-bitter acids. Thus the old hypothesis concerning the biosynthetic relationship between myrcene and α- and β-bitter acids has been verified. Mutual relations continue during the storage of hop cones when myrcene continues to stimulate the oxidation of α-bitter acids. Therefore, hops with more myrcene, e.g. American varieties, are more rapidly transformed during storage and become old sooner than Continental European varieties.

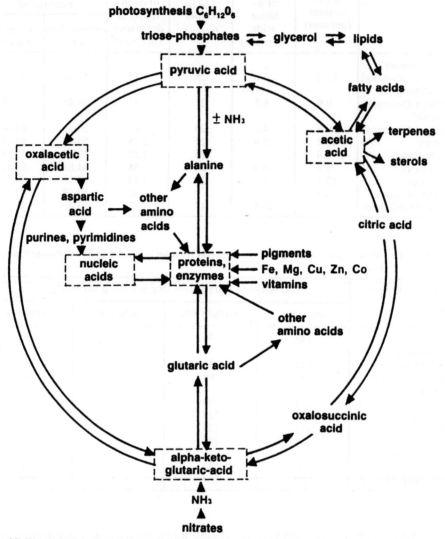

Fig. 29. Biosynthesis of terpenes and lipids.

Hop tannin-polyphenol substances

According to SALAČ (1954) the hop tannin was previously regarded as harmful in the brewing of beer. But SALAČ (1954) and his team showed that hop tannin is a very efficient component which is typical of Czechoslovak hops. POZDĚNA and JERMOLAJEV (1962) found tannin in underground organs as well as in ripening cones, but not in other above--ground parts.

As well as its value for brewing hop tannin is also very important as an indicator of the quality and origin of hops as it has a definite relationship with hop resins. HUBÁČEK (1964) elaborated a chemotaxonomy for Czechoslovak hops according to differences in the polyphenol substances (tannin) in ripe cones. The content of particular components of hop tannin depends on the inherent properties of the variety as well as on the external ecological conditions. The content of tannin-polyphenol substances in different parts of seedless hop cones is shown in Table 19. Unlike the content of resins, the content of tannin decreases during the ripening of cones.

TABLE 19 **Content of tannins (polyphenol substances) in different parts of the hop cone**

Component part	Savov and Kovačev (1953)		Bulgakov (1954)	Wollrab (1957)	Dyr (1962)
	tannin (g per 100 g dry matter)	relative content of tannin			
Lupuline glands	0.300	7.9	6 – 9	22	22
Bract	3.000	78.9	90	76	74
Cranked axis	0.275	7.3	1 – 4	2	4
Rachis	0.225	5.9	0	not given	not given
Total	3.800	100.0	100	100	100

A comparison of the amounts of resins and tannin in different parts of the hop cone shows that the content of resins is highest in lupuline glands and that of tannin is highest in bracts, whereas both are lowest in the rachis and cranked axis. Differences in the quantity of rachides between manually and machine picked hops will thus affect the amount of resins and tannins in the harvested cones.

The hop tannin (polyphenol substances) is divided into five fractions:
1. flavonol glycosides,
2. aromatic oxyacids and methoxyacids,
3. anthocyanins and their leuco-forms,
4. coumarins and their glycosides,
5. free acidic phenols.

Many compounds have been identified in each of the fractions but it is still assumed that only relatively few of these substances are yet known. The chemistry of tannin components is, as yet, only little known due to their great complexity. So, knowledge about the polymerization of polyphenol substances is based to some extent on experimental studies but also, however, on hypothetical postulates.

It has been found that the ratio of the individual substances within the group changes the efficiency of both hop tannin and the essential oils and resins. It has also been found that the total efficiency is affected by the relationship between particular groups of the major active substances. MOŠTEK (1969) found that hop tannin influences the transformation of α-bitter acids especially that of humulone by its dehydratation effects. If tannin is added to a boiling aqueous solution of humulone, there is a rapid increase in the intensity of bitterness. According to MOŠTEK (1969) the content of bitter substances falls by 15 per cent (including a fall in the iso-humulones by 20 per cent) and, in contrast, the content of nitrogenous substances increases by 15 per cent if one component of tannin anthocyaninogene is reduced in the boiling beer. This implies that the mutual effect concern not only the active substances but also the accompanying substances which are present in hop cones and can exercise a positive or negative influence on the active substances.

The dry matter of hop cones contains about 20 per cent active substances, i.e. essential oils, resins and tannins. Another 2–3 per cent includes hormones and pharmacologically active vitamins among them thiamine, nicotinic acid, panthotenic acid, biotine, pyridoxine and oestrogen. The rest of the dry matter consists of 5 major groups of a c c o m p a n y i n g s u b s t a n c e s :

1. saccharides (monosaccharides, oligosaccharides; pectin, hemicellulose, cellulose);
2. organic acids;
3. fats and waxes;
4. nitrogenous substances;
5. mineral substances (ash).

Of these the saccharides are the most common representing up to 50 per cent of the dry matter. In analysis they are divided into two groups: cellulose at about 15 per cent of dry matter and extracted nitrogen-free substances at about 35 per cent. In this classification the nitrogen-free extractive substances include the essential oils and tannin, some of them combined with saccharides to form glycosides. But the major representative in this group is pectin (12–16 per cent of dry matter) with positive effect on the active substances due to its function as so-called protective colloids.

There are numerous sugars in hop dry matter (monosaccharides and oligosaccharides). MAC WILLIAM (1952) found 0.38–0.55 per cent of fructose, 0.32–0.44 per cent of glucose and 0.10–0.57 per cent saccharose as well as small amounts of raffinose, stachyose and some pentosans. The analyses of RYBÁČEK (1963) found a total of 3 per cent of reducing substances mainly reducing monosacchárides in the dry matter from cones and a previously unmentioned content of starch (0.03 per cent). Starch has been found in all of the above-ground organs of hop plants.

The organic acids, succinic, malic and citric have been found in fresh hop cones. Moreover, isovaleric and butyric acid was found in hops stored for a long time.

The important nitrogen-free extractive substances found in hop cones are shown in Table 20.

Other nitrogen-free substances have also been isolated, including the wax, myricine at 0.25 per cent, ceryl alcohol, phytosterol and palmitic and stearic acid, the oxidation products of which contribute to the rancid smell of old hop cones.

TABLE 20 **Nitrogen-free extractive substances in hop cones**

Material group	Dry matter content (per cent) according to			
	Savov and Kovačev (1953)	Dyr and Hauzar (1962)	Rybáček (1963)	Burgess (1964)
Total nitrogen--free extractive substances	–	–	34.47	–
Tannin	3.53	2.28 – 5.71	4.57	4.00
Pectins	13.71	13.71 – 16.00	14.95	12.00
Reducing sugars	4.00	4.00	3.00	2.00
Starch	–	–	0.03	–
Other nitrogen--free extractive substances	–	–	14.92	–

There are many nitrogenous substances in hop cones (10–24 per cent of dry matter) but they have not yet been intensively investigated. According to BULGAKOV (1954), asparagine makes up about 30 per cent of all nitrogenous substances. Among the other low-molecular weight, nitrogenous compounds present in hop cones, he mentions histidine, arginine, betaine, adenine and hypoxanthine. He also found albumins, peptones and polypeptides.

Hop cones contain 6–10 per cent of mineral substances (ash) in their dry matter. The individual elements are mentioned in Table 13, which shows the high representation of potassium (K), calcium (Ca), phosphorus (P) and magnesium (Mg).

The fruit of the hop plant – achene (stone)

The fruit of the hop plant is one-seeded achene (in hop-growing terminology it is called a stone). But the presence of fully grown fruits reduces the quality of hop cones, therefore most attention has been paid to the comparison of fertilized and un-fertilized (seeded and seedless) cones, so that the fruits themselves have had relatively little left for investigation. In the period of technical maturity, hop cones contain no seeded fruits and in technically ripe, but lowgrade cones the seeds are not fully grown.

Chemical analysis of seedless cones and cones with seedbearing fruits has shown that those with seeds had more fat and proteins.

The fruits of hop with physiologically ripe seeds were analyzed by RYBÁČEK (1963) and his results are shown in Tables 12 and 13.

These tables show that hop fruits contain more nitrogenous substances and ether extract (oil) and less nitrogen-free extractive substances, cellulose and ash than do the cones. The precursor for the production of fatty acids and hop resins is the acetic acid, as shown in Fig. 29. This fact makes it clear, that the quality of seeded cones is lower than in the seedless ones.

There may be up to 25 per cent of fruits in dry cones of some foreign hops. ROBERTS and STEVENS (1963) found 32 per cent of saponifiable and 2.2 per cent of unsaponifiable fraction, containing beta-sitosterole, in hop fruits. The saponifiable fraction contained fatty acids in the form of methyl esters (linoleate – 60 per cent, linolenate – 15 per cent, oleate – 10 per cent, palmitole – 10 per cent and stearate – 3 per cent).

Fig. 30. Fruit – achene: 1 – achene, 2 – seed, 3 – cross section of seed (a – seed coat-testa, b – hyaline layer under the seed coat, c – foetus-embryo, d – endosperm).

PHYLOGENESIS OF THE HOP

It is necessary to study the historical development of the h o p g e n u s (*Humulus* L.) and its most important species, the c l i m b i n g h o p (*Humulus lupulus* L.) in order to obtain knowledge of general genetic properties of the species. Genetics is the starting point for hop cultivation and is necessary to preserve the contemporary varieties and to produce new, improved strains.

Many of the biological properties which have developed in the long phylogenesis of the climbing hop have been exploited as economically important and profitable. The genetic properties of the hop constitute a firm base for its cultivation, but the plant also possesses certain other factors of non-genetic variability. This variability is manifested during the annual cycle of ontogenesis and, together with the fundamental genetic properties provides a biological base for agrotechnology.

According to botanical systematics, the whole phylogenesis can be divided into a number of parts involving that of the genus, its species and subspecies, and that of the wild and cultivated variety of European hop and its subvarieties.

Phylogenesis and systematics of the hop genus (*Humulus* L.)

If the genetic phylogenesis is understood, the knowledge can be used for the systematics both of the genus, its species and subspecies.

However, the systematics of hop are not yet fully stabilized because the genus has not been firmly fixed in a particular family and it has not yet been clearly separated into individual species. LINNÉ (quoted according to STOCKER 1963) placed the genus hop in the family of *Moraceae*. ŽUKOVSKIJ (1938) like many other authors also placed hop in the *Moraceae*. GALKA and NADAŠKEVIČ (1927, quoted by NEČIPORČUK 1955) put it in the Urticaceae. KAVINA (1923), TISCHLER (1927), DOSTÁL J. (1954) and other authors concluded that it should be in the *Cannabaceae*. However, VENT (1963) suggested that there are good reasons for excluding *Humulus* L. from the *Cannabaceae* and for forming an independent family. If this is not to happen, then it would seem that including *Humulus* L. in the *Cannabaceae* is the most suitable solution.

There are five independent genera of hop described in the available literature, though some of the sources entertain serious doubts about two of them. At present there are no objections against the independence of the remaining three genera. The p e r e n n i a l c l i m b i n g h o p (*Humulus lupulus* L.) with 20 chromosomes is the best known genus. The cultivated varieties, differing in certain features and properties, have been developed from this genus. Variability in characteristics and properties occurs widely in wild plants.

A more sparsely leaved wild hop was discovered in North America in 1847, claimed by NUTALL to be an independent genus and designated as the A m e r i c a n h o p (*Humulus americanus* Nutall). However, the generic independence of this hop was later considered to be very doubtful. Thus, FORE and SATHER (1947) claimed this genus to be identical with the hop which grew wild in the North American state of New Mexico and is classified as *Humulus lupulus* L. var. *neomexicana* Nels. et Cockerell. DARK (1951) did not recognize the generic indepencence of the American hop and assumed it to be the climbing hop. We accept this view, because the quoted characteristics and properties involving a different scent of the cones and a different density of leaf-clad cannot be sufficient to confirm generic independence.

Recently the generic independence of the heart-leaf hop was also put in doubt. In the past this hop has been recognized as an independent genus *Humulus cordifolius* Mig. It grows wild in Siberia, in the mountainous areas of Altai and in the Far East.

Both of the formerly indenpendent genera (*Humulus cordifolius* Mig. and *Humulus neomexicanus* Nels. et Cockerell) have the same number of somatic chromosomes (2n = 20), but the sexual chromosomes of the male plants have typical deviations as can be seen in Fig. 31. Their individual populations occupy certain areas of the locality inhabited by the whole genus and where the areas meet, there are transient forms thus satisfying the criteria for the determination of a subspecies. According to RYBÁČEK (1978) they can rightly be taken as the independent subspecies *cordifolius* Maxim. and subspecies *neomexicanus* Nels. et. Cockerell.

Species	Humulus lupulus					Humulus japonicus
Subspecies	neomexicanus		europaeus	cordifolius		
PMC — x chromosome	x	x	x	$x_1\;x_2$	$x_1\;x_2\;x_3$	x
meiosis — y chromosome	y	y	y	$y_1\;y_2$	$y_1\;y_2\;y_3$	$y_1\;y_2$
Dimensions	10:10	10:8	10:5,2	10:12:14,2:7,2 10:12,7:10,9:3,4		
Type of sexual chromosome	homotype	"New Winge" type	"Winge" type	"Sinoto" and "New Sinoto" type	hexapartite complex	tripartite complex

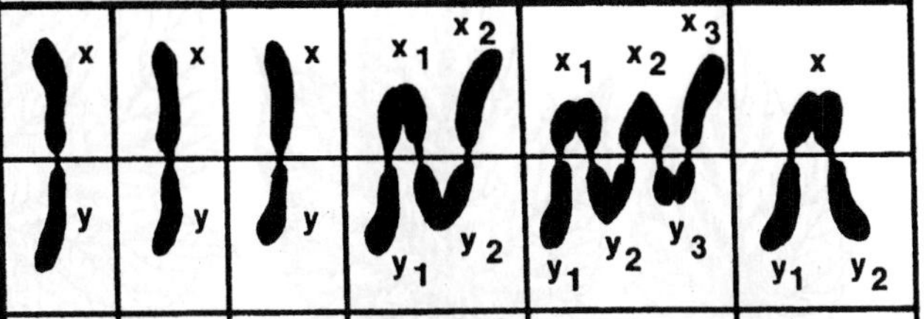

Fig. 31. Sexual chromosomes in male plants of climbing and Japanese hop.

Other populations of the genus *Humulus lupulus* L. meet the required criteria of a subspecies and therefore RYBÁČEK (1978) proposes to designate the one found in Europe as subspecies *europaeus* Ryb. Independent subspecies can be distinguished by the formation of joined sexual chromosomes (x. y) of the male plants as shown in Fig. 31. The varieties "Simoto" and "New Simoto" with joined sexual chromosomes in the male plants are *Humulus lupulus* L., subspecies *cordifolius* Maxim. "Winge" occurs in male plants of the European hop and the most simple bivalent homotype with joined sexual chromosomes "New Winge", is the New Mexican hop. The "New Winge" type which is different from the homotype "Winge", is probably a cross between the wild New Mexican hop and the wild European hop because both are present in the natural stocks of North America.

These three subspecies can be distinguished morphologically, by the form of their leaves and cones.

The h e a r t - l e a v e d h o p (*Humulus lupulus* L., subspecies *cordifolius* Maxim.) has, as its name implies, leaves which are compact, unlobed and heart-shaped. Its cones are small with a small amount of lupuline.

T h e E u r o p e a n h o p (*Humulus Lupulus* L., subspecies *europaeus* Ryb.) is distinguished morphologically from the heart-leaved hop by having heart-shaped leaves only at the tips of bines and side branches. All other leaves are lobed with three or five lobes. Its cones are bigger and have a greater content of lupuline.

T h e N e w M e x i c a n h o p (*Humulus lupulus* L., subspecies *neomexicanus* Nels. et Cockerell.) is less leafy and has heart-shaped leaves only at the tips of fruiting branches. Other leaves are lobed, even those on the upper third of the bine. Its cones are big with a greater content of lupuline but with a non-typical beer scent.

The first leaves on young seedling plants of all three subspecies are entire and cordate. The existence of lobed leaves can thus be taken as a secondary phenomenon which appears on phylogenetically younger populations. Therefore, the heart-leaved hop can be regarded as phylogenetically the oldest and the New Mexican hop as the youngest.

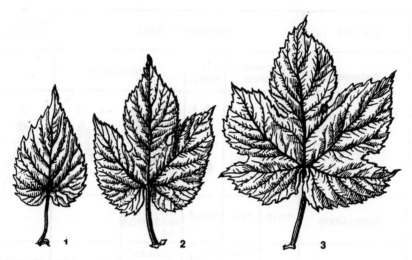

Fig. 32. Leaves of climbing hop (*Humulus lupulus* L.): 1 – cordate leaf, 2 – leaf with three lobes, 3 – leaf with five lobes.

Fig. 33. Leaves of Japanese hop (*Humulus japonicus* Sieb. et Zucc.): 1 – leaf with five lobes, 2 – leaf with seven lobes, 3 – leaf with nine lobes.

As well as the c l i m b i n g h o p (*Humulus lupulus* L.) which is a perennial plant, two other species have been identified and these are annual plants. One of them is the J a p a n e s e h o p (Humulus japonicus Sieb. and Zucc.), whose female plants have 2n = 16 somatical chromosomes and male plants 2n = 17. These different numbers of chromosomes in male and female plants are caused by the trivalent joined sexual chromosomes of the male plants as discovered by ONO (1955). The annual plants of Japanese hop are propagated exclusively by seeds. The plant has leaves with five to nine lobes, its cones are small with stiff bracts and contain only a small amount of lupuline. In nature, this hop is found in north-east China (Manshuria), South Korea and Japan.

Fig. 34. Japanese hop (*Humulus japonicus* Sieb. et Zucc.): 1 – female plant, 2 – cone.

The twining hop (*Humulus scandens* Lour. et Merrill.) is also an annual plant but with entire cordate leaves. It is found wild in Central Asia. According to WINGE (1950) it is distinguished from the climbing hop by a different number of chromosomes in its somatic cells.

Phylogenesis and systematics of the European hop (*Humulus lupulus* LINNÉ subspecies *europaeus* RYBÁČEK)

Three varieties of European hop can be distinguished, as follows:
1. stunted (varietas *irenae minima* BLATT.);
2. wild (varietas *spontanea* Ryb.) and
3. cultivated (varietas *culta* Ryb.).

The criteria for distinguishing the wild and cultivated European hop were given precision by RYBÁČEK (1978).

European dwarfish hop was discovered by BLATTNÝ (1963) in Czechoslovakia and Rumania. It has small leaves and cones the size which does not increase when the plant is given better growing conditions. In the nature it is found in mountainous localities, most commonly in the Carpathians. Its propagation is usually vegetative.

The European wild hop is widespread in the temperate zone of the northern hemisphere, especially around the 50th parallel. It is most common in lowlands but occurs also in highlands and the foot-hills of mountains. It features both vegetative and generative propagation and this has contributed to its expansion as well as variability of its marks and properties. Its cones are rough, non-cultivated and useless for brewing as they provide only bad flavours.

The European cultivated hop developed from the wild hop by gradual domestication and later by conscious cultivation. It is maintained by the vegetative propagation of cultivated stocks in hop gardens and of wild stocks being kept for use in breeding. If the European cultivated hop is allowed to propagate by seed it reverts to the wild variety. The basic difference between the wild and cultivated varieties of European hop lies mainly in the genetically fixed higher uniformity in blossoming and in the ripening of cones.

Phylogenesis of the European wild hop

As is evident from the map in Fig. 35, the European wild hop considerably extended its original locality. This expansion has undoubtedly been aided by the semi-cultivated and cultivated varieties which have become wild in the territories of its cultivation. Some authors, including BLATTNÝ (1950) do not consider the hop to be an autochthonous European plant. They consider that the stocks of wild hop originated from semi-cultivated and cultivated hops which became wild after they had been brought to Europe. Even with vegetative propagation there is a certain regression of the cultivated varieties of hop towards the natural state, demonstrated for example by a prolongation of the vegetative period and an increasing diversity of the blossoming and ripening periods. Therefore uncultivated stocks containing male and female plants can, justifiably, be assumed to be wild stocks. Both males and females of hybrid strains grow from seeds and the strains are distinguished by the different colours of the bines. *The red bines* have reddish, red or dark violet bine due to an anthocyanin pigmentation. This pigmentation is missing from the *green bines*. The populations of both, red bines and green bines, are assumed to be subvarieties. These subvarieties of European hops are ecologically distinguished by their resistance to frost and by their requirements for humidity. The green bines, without anthocyanin cannot withstand severe frosts and require

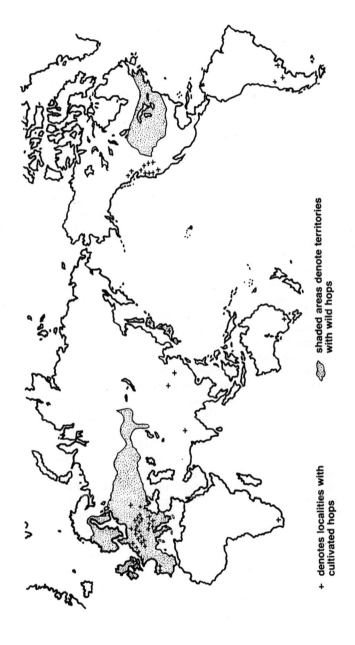

+ denotes localities with
cultivated hops

shaded areas denote territories
with wild hops

Fig. 35. Geographical distribution of European wild and cultivated hops (according to BURGESS).

more water. The red bines evidently represent a phylogenetically older subvariety and are of continental origin. It is accidental that the older European varieties, like 'Žatecký', 'Hallertauer' and the English 'Golding' are red bines. The green bines are better able to accommodate to maritime conditions. This accommodation could have developed in territories around the Caucasus and Black Sea where wild hops are widespread. Green bines as well as red bines are also found in other large areas of the Soviet Union. Many natural localities in Czechoslovakia have red bines as well as green bines of wild hop. The best conditions for the wild hop are found in warm and humid localities like flooded forests, valleys of streams and rivers, around fishponds, in dingles, on hillsides and in any other places well provided with humidity, from low land up to high land. It is most common in stands of leaf-bearing trees and bushes where is a relatively high level of soil moisture. It does not inhabit coniferous forests nor dry localities.

Phylogenesis of European cultivated hop

The phylogenesis of the cultivated hop began with the domestication of the wild climbing hop. Documents relating to the early period of the domestication and cultivation of wild hops are very rare. There were difficulties with the culture of the original wild hop from the Altai-Mountains, as mentioned by GARBUZOVA (1956), and it is likely that before cultivation began, the wild hop underwent certain changes in the neighbourhood of human settlements. The influence of these settlements, enriched with organic nitrogenous substances, served as an efficient cultivating intervention during this period. Hence the hop plant became adapted to a greater abundance of nitrogenous substances from the environment, so that the mycorrhiza normally present on the roots of wild hops became less important in nutrition. Another great intervention was a decapitation. This had certain unfavourable physiological consequences, as mentioned in the literature (e.g. RYBÁČEK, 1960), but it was a spontaneous selective intervention which led to the production of populations of early hop. The main aim of decapitation however, was to facilitate the picking of the cones of wild hops climbing on broadleaved trees and bushes.

Real cultivation began with the conscious vegetative propagation of hop plants. This made it possible to transfer hop to fenced gardens and to produce smaller, homogenized populations of the cultivated hop. When the hop was cultivated in gardens it required supports around which it could climb. According to early reports hop growers imitated nature by growing the plants near to dried trees and shrubs. Later, they used wooden supports. This led to another cultivating intervention namely that of hop tying. The importance of this does not lie solely in the tying of bines to their supports, but also in the necessity to reduce the number of bines and to select those suitable for tying. Selective spring hoeing had a similar function, in that it accompanied the selection of the best-grown bines before they were tied. These two interventions encouraged earlier flowering and, together with early plants in local populations of Czechoslovak hops. This explains why the original district varieties like 'Staročeský červeňák' ('Old Czech red bine'), 'Starožatecký červeňák' ('Old red bine from the region around town Žatec') and 'Staroúštěcký červeňák' ('Old red bine from the region around the town of Úštěk') were early varieties. This high degree of earliness was positively reflected in a higher quality of cones and their greater homogeneity. On the other hand, the lower yield represented a negative feature. Therefore the first experiments into delayed growth of hop plants began, in Czechoslovakia, with the aim of reducing the unfavourable consequences of excess earliness.

According to MOHL (1906) deep hoeing in the spring was first applied in the region of Úštěk. It produced higher yields and better quality cones. This very effective agrotechnical measure is now widely used and has been experimentally verified in a number of different variants. In the search for delayed flowering, half-early plants of the variety 'Staroúštěcký červeňák' were selected in hop stocks and further propagated. It was Mr Kryštof SEMŠ, a hop grower from the town of Úštěk, who succeeded in this conscious breeding activity when, in 1856, his selections produced an excellent half-early variety of red bines, given the name 'Semšův chmel' (i.e. 'Semš's variety'). This variety represented a great contribution to the local crop and later to the whole of Czechoslovak hop growing.

Phylogenesis of varieties and subvarieties

Commercial varieties represent the results of the later stages of phylogenesis. Older varieties resulted from nonintentional (spontaneous) breeding, but later the selection breeding and cultivation was intentional.

All varieties originate from genetically heterozygous individuals, and from a great number of such individuals where populations are concerned or from one plant in the case of clones. The properties of varieties are affected by the length of time over which they are propagated in one district. With long-term vegetative propagation of hop, as occurred with the original local varieties, a greater number of bud mutations is likely to occur, from which new varieties can be obtained by selection.

According to the method of selection that was applied the following types of commercial varieties may be distinguished:

1. the original local varieties, which appeared in a particular district;

2. the cultivated local varieties, which arose from the local variety after repeated negative selection, but with such a relatively broad genetic basic that their properties are not very different from the original local variety;

3. collective selections from a local variety, obtained by positive selection of individuals; this has a limited genetic base proportional to the number of individuals included in the selection;

4. selected clones; these are the vegetatively propagated progeny of a selected individual, hence they have the most limited genetic base;

5. hybrid clones obtained through the crossing of selected suitable individuals and with the subsequent selection of the best seedlings for further vegetative propagation (hybridization makes it possible to extend the genetic base of the clone).

According to the length of the period of vegetation, there are early, half-early, half-late, late and very late varieties. VENT (1963) considered early and late maturity of both red and green bines, to be closely correlated with the speed of growth of cones, as indicated by the thickening of their bracts and bracteoles. In early and half-early varieties this thickening is gradual and lasts 10–14 days whilst the thickening of half-late and late varieties lasts only about seven days. Nevertheless the vegetative period of late varieties is longer, blossoming is later, and the flowering and ripening periods are longer.

According to their cone quality there are aromatic (fine or cultivated) and other (rough) varieties. The quality and degree of fineness depends also on the length of phylogenesis which is shown by comparing the length of phylogenesis of higher and lower systematic botanical units (species, subspecies, variety and subvariety) and also by knowledge of the phylogenesis of varieties. Therefore, knowledge of the phylogenesis of varieties is important not only for cultivation and breeding but also for agricultural technology.

According to their genetic origin it is possible to distinguish genetic groups which link the varieties of the European hop. These groups include genetically related varieties which may have different territorial origins. Long-term breeding, and especially controlled crossing can produce hybrid clones, whose cone quality approaches that of the original genetic group but at the same time has introduced new desirable qualities such as higher disease resistance.

71

In considering the phylogenesis of varieties, most attention will be given to the genetic group of "Žatecký červeňák" and to some of the most important representatives of other genetic groups.

The first genetic group "Žatecký červeňák" is assumed to be the oldest group of varieties. It originated from the old variety 'Staročeský červeňák' (i.e. 'Old Cech red bine') from which the varieties 'Staročeský červeňák' (i.e. 'Old red bine from the town and region of Žatec') and 'Staroúštěcký červeňák' (i.e. 'Old red bine from the region of Úštěk') originated. As already mentioned, Kryštof SEMŠ produced the half-early variety of red bine called 'Semšův chmel' (i.e. 'Semš's variety', 'Semš's hop') and this variety because of its excellent commercial properties became established in all of the Czechoslovak hop regions. In some localities 'Semš's hop' was cultived along with 'Starožatecký červeňák' and 'Staroúštěcký červeňák'.

According to OSVALD (1944) the original variety 'ŽATECKÝ' (i.e. 'Hop from the town and region of Žatec') arose from mixed stocks. Hence this is the single variety (cultivar) cultivated in all Czech regions and therefore where Czech hop-growing literature mentions varieties they are not varieties in the true sense but only local modifications (modification from the town od Žatec, Úštěk, Tršice etc.) Because the Czechoslovak list of authorized varieties quotes only independent commercial varieties, all local modifications are listed as varieties in spite of the fact that they are only parts of one variety, i.e. they are subvarieties.

The variety (more accurately, the subvariety) 'Úštěcký krajový' gave rise to a new cultivated regional variety of half-early red bine called 'Blato' (i.e. 'Region of Blato'). Similarly the variety 'Žatecký krajový' produced a regional variety of half-early red bine called 'Lúčan'. The advantage of original varieties can thus be seen to be their broad genetic base which has the potential to provide new strains with a greater stability of quality and yield even though their own yield may be somewhat lower.

This broad genetic base can be preserved in the synthetic populations selected from regional

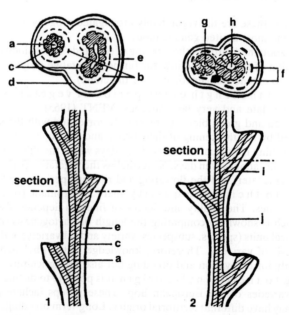

Fig. 36. Cranked axis (strig) of cones: 1 – red bine, 2 – green bine, a – pith, b – rings of phloem in vascular bundles, c – xylem cylinder of vascular bundles, d – epidermis, e – cortical tissue, f – sclerenchyma, g – main axis, h – lateral branching to bracts, bracteoles and fruit, i – small side branch, j – axis of strig.

varieties. Thus Výzkumný ústav chmelařský (the Research Institute for Hop Growing) in Žatec succeeded in producing from the variety 'Žatecký krajový' two authorized subvarieties – 'Aromat' and 'Siřem'.

OSVALD cultivated selected clones by the propagation of selected individual plants from stocks of the variety 'Žatecký krajový' planted in the Žatec region. Three of them were recognized in 1946 as independent varieties and were designated as 'Osvald's clone No. 31', 'Osvald's clone No. 72' and 'Osvald's clone No. 144'.

By a similar procedure the Research Institute of Hop Growing in Žatec cultivated a clone designated as 'Zlatan' (i.e. 'Golden') from the variety 'Žatecký krajový' and this is now recognized as an independent variety.

The excellent quality of Czech regional varieties and 'Semš's hops' caused their cultivation to spread outside Czechoslovakia. Many of the important varieties and subvarieties found over most of Europe originate from them. In the Federal Republic of Germany one such important early variety is 'Tettnang', cultivated in the Tettnang hop-growing region as is 'Spalt' a half-early red bine cultivated in the Spalt area.

Very many varieties derived from the genetic group of red bines from Žatec have appeared in the Soviet Union, where the variety 'Semš's hop' is preserved, by name, in the Ukraine, where it was imported in 1890. From 'Semš's hop' and other imported Czech regional varieties, new local regional varieties appeared in the region of Žitomir and Volyň as e.g. the half-early red-bine varieties 'Ivanovickij' (i.e. 'Hop from Ivanovice'), 'Volyňskij' (i.e. 'Hop from the region of Volynia') and 'Rogatynskij' (i.e. 'Hop from the region of Rogatyn'). The research station in Žitomir produced from 'Volyňskij' a clone called 'Žitomir No. 18' which became widespread in the region of Žitomir. At this time clones 'Žitomir No. 26' and 'Žitomir No. 34' were also developed.

In Poland Czech hops served as the base for the cultivation of the of half-early red bine 'Opolski' (i.e. 'Hop from the region of Opole'). More recently two new varieties have appeared, namely 'Lubelski' (i.e. 'Hop from the region of Lublin') and 'Nadwislanski' (i.e. 'Hop from the region around the river Wisla').

The second genetic group of "German red-bines" includes the varieties 'Hersburger' (early red-bine cultivated in the Hersburg-hills), 'Hallertauer' (half-early red-bine which appeared first in the Bavarian hop-growing region of Hallertau) and the more recently cultivated 'Hallertauer Gold'. 'Hüller Anfang' and 'Hüller Start'.

The third genetic group of English red-bines known as "Golding" includes many early, half-early and late varieties. Their origin was a half-late variety 'Old Golding' derived, by selection, from an even older late variety 'Canterbury Whitebines' as early as 1790. All varieties of this genetic group are designated, according to English usage, with the mark "G".

The variety 'Old Golding' gave rise by selection to the early clone 'Bramling' (G) and this produced an even earlier clone 'Amos Early Bird' (G). Among the half-early varieties of the "Golding" group the most renowned are 'Eastwell' (G), 'Gobles' (G), 'Petham' (G), 'Rodmerham' (G), 'Tutsham' (G) and the more recent hybrid variety 'Whitbread 1147' (G). Varieties of the "Golding" group were involved in its development. This is evident from its quality which approaches the quality of these varieties. This variety possesses a high level of resistance to the dangerous Verticillium wilt.

The "Golding" group found application in countries other than England. In Yugoslavia the variety 'Savinski Golding' is in third place after the North Bohemian (Žatec) and the West German (Hallertau) red bines.

Other hybrid red bine varieties do not form a genetically homogeneous group, though they possess certain compatible components useful in crosses. The oldest is the Soviet half-late variety 'Serebrjanka' (i.e. 'Silver hop') which occurred as a wild seedling. It has a dark-red to violet bine and high degree of resistance to Peronospora. This property led to its use in a cross which produced the new American variety 'Cascade'.

Another widespread variety is the English hybrid red bine 'Northern Brewer' which has a high level of resins (18–20 per cent or even more).

New hybrid red-bines were developed in England (the year of the authorization is given in brackets), namely: 'Wye Northdown' (1970), 'Wye Challenger' (1972), 'Wye Target' (1972), 'Wye Saxon' (1973), 'Wye Viking' (1973), 'Yeoman' (1982), 'Zenith' (1982) and 'Omega' (1984). In Yugoslavia the hybrid subspecies 'Dunav' (1972), 'Vojvodina' (1972), 'Atlas' (1976), 'Apolon' (1976), 'Bobek' (1981) and 'Buket' (1981) were developed.

The genetic group of "French-Belgian green bines" is probably the oldest among the green bines. In France they are well represented by the late varieties 'Vert d'Alsace' and 'Petit vert de Lucey' and the half-early 'Precoce de Dienlorande' and, in Belgium, by the half late 'Tige Verte', 'Buvrinnes' and the late varieties 'Cono' and 'Rusbus'.

A separate *genetic family of green bines* is formed by *"Bač green bines"* cultivated in the Bačka region of Yugoslavia. They most probably originate from the variety 'Dubský zeleňák' (i.e. 'Green bine from Dub'), formerly cultivated in Czechoslovakia.

"Fuggles" is a rich genetic group of *English green bines.* They were derived from the variety 'Fuggles' produced in 1875 by the free pollination of the mother variety 'Colgat'. Many clones were selected from this variety, and 'Fuggles N' was one of the best of them.

American green bines, with the name *"Cluster",* constitute an independent genetic group. They probably originated from the variety 'English Cluster'. The most wide-spread of them is a late green bine called 'Late Cluster', which has sometime been designated according to the area of its cultivation, e.g. 'Oregon Cluster'. There is a half-early green bine named 'Early Cluster'.

The rest of the hybrid varieties of green bines do not form a genetically homogenous group although some of them have one common parent. Two late green bines which possess genes for high yield and content, known as 'Bullion Hop' and 'Brewers Gold' are sister seedlings justifiably taken to be independent hybrid varieties due to their different features and properties, produced by different genetic combinations.

From the Federal Republic of Germany new important hybrid varieties should be mentioned as follows: 'Hollertauer Gold' (1974), 'Hüller Bitterer' (1974), 'Perle' (1978) and 'Orion' (1984). From the United States the varieties 'Columbia' (1976), 'Willamette' (1976) and 'Comet' (1974).

Another group of varieties containing more than 10 per cent α- acids comes from the United States. This group is designated as *"Superalpha group"* and includes the following varieties: 'Galena' (1978), 'Eroica' (1980), 'Nugget' (1983), 'Olympic' (1983) and 'Chinook' (1985).

Varieties authorized for cultivation in Czechoslovakia

In 1978 eight varieties were authorized for cultivation in Czechoslovakia. Five major groups were described above and from these groups the authorized varieties depending on the cultivation methods and their properties, are separated into three different types, namely cultivated regional varieties, collective selections and selected clones.

All authorized Czechoslovak varieties, though different in cultivation type belong to the fine quality half-early hop from the genetic group "Žatecký červeňák". Therefore they bear cones of high quality with no marked differences in their internal and external qualitative features. There are no difference at all in the final product i.e. in the dry cones, a fact of great importance for the qualitative uniformity of hops throughout the hop-growing regions of Czechoslovakia. Likewise no big differences are to be found in the features of the above-ground organs of the authorized varieties. However varieties differ in the size of their genetic base, which is most compact in individual clones and broad in the cultivated regional varieties as would be expected from the breeding methods used in their development. This, then also implies certain biological dissimilarities, especially small differences in the length of vegetative period and in the argo-economical requirements of individual varieties.

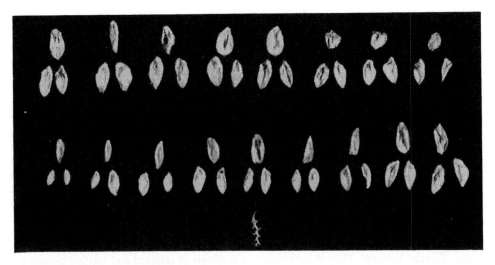

Fig. 37. Hop cone of high quality in analysis.

The oldest among the authorized varieties are two cultivated regional varieties 'Lúčan' and 'Blato' obtained by a systematic repeated negative selection within chosen stocks of the original regional varieties. This cultivation procedure preserved the very broad genetic base of the original regional varieties. Present populations of these varieties consist of large numbers of partially different groups, differing only in certain physiological and ecological properties. They represent a mixture of the biotypes of those populations, including the original hops cultivated in different regions of Czechoslovakia, which produced the morphologically unified variety 'Žatecký'. This is the origin of their high adaptability (plasticity) to different external conditions (especially climatic) as well as of the more stable quality and yield of cones which is accompanied, however, by a lower average yield.

The variety ' L ú č a n ' was obtained from the variety 'Žatecký krajový' (registered 1941) after its trueness to type had been confirmed and its propagation improved. It is a half-early red bine of medium height with a cylindrical habit. The bine is reddish. If it is grown in a confined space then its fertile shoots occur of 150–160 cm above ground level, but if there is plenty of room the corresponding height is 230–260 cm. The inflorescence is 60–100 cm long with a semipropendent upper half and a propendent lower part. The cones are egg-shaped and

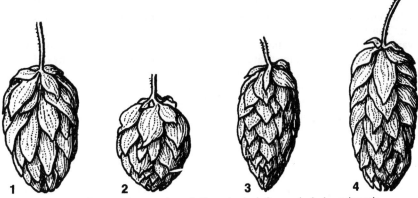

Fig. 38. The shape of cones: 1 – egg-shaped, 2 – spherical, 3 – conical, 4 – prismatic.

2–3 cm long. Their structure as well as the cranking of the axis are symmetrical. Both fresh and dried cones have the characteristic very fine hop scent. Their vegetative period lasts 122–127 days. 'Lúčan' is a reliable variety in the Žatec hop region. Having a broad genetic basis means that it has no special requirement for locality and can pick up with humous soils. In a confined locality the variety is more disposed to infection by *Peronospora*.

The variety 'B l a t o' (registered in 1974) originated from the regional variety 'Úštěcký krajový' (registered in 1952) by the same breeding procedure used for the variety 'Lúčan'. It is a half-early red bine of medium to large height and with a cylindrical habit. The colour of the bine is reddish. The height at which fertile shoots are produced is 140–170 cm in a confined space and 220–250 cm with ample space. The fertile shoots are 50–90 cm long. The upper half of the bine is semipropendent and the lower half is propendent. The structure of cone and the cranking of the axis are both regular. The cones have a true hop scent. The variety ripens 3–5 days later than 'Lúčan' in the Žatec region. The variety 'Blato' is suitable for all Czechoslovak hop-growing regions and has no special requirement for locality. Because it is resistant to Peronospora it can be used in more humid areas.

The two varieties 'Aromat' and 'Siřem' are also in the group of selected populations. They were produced by plants individually selected from population stocks of the half-early red bine 'Žatecký krajový'. They thus represent a synthesis of selected and verified clones, the mixture of which has produced these new synthetic populations. The yield of cones of these varieties equals that of the standard variety 'Osvald's clone No. 72', but they contain a little more of the resins, especially the α-component (humulone). Their broad genetical base is reflected in high adaptibility of both varieties, making them suitable for use in all Czech hop-growing regions.

The variety 'A r o m a t' was bred by three-phase selection from the half-early red bine "Žatecký krajový" taken from stocks in Lhota pod Džbánem in the Podlesí region. This selection propagated in the nursery of the Research Institute for Hop Growing in Žatec gave 1100 plants to provide a synthetic population of this variety. Its above-ground parts are those of a half early red bine. It is of medium height reaching a little over the top of the hop poles at 7 meters. Its shape is regular and cylindrical. When grown at narrow spacing the fertile shoots are produced at a height of 150–180 cm, and with wide spacing the figure is 230–260 cm. The bine is green red in colour with middle-sized leaves slightly inclined. The bine is on average 10 mm in diameter and the length of the internodes is 27–30 cm. It is a good climber. The shoots in the lower part of the bine are propendent, and 90–115 cm long, whilst in the upper part they are semipropendent and 50–110 cm long. The cones are a slightly elongated oviform with a regular symmetrical structure and with bracts and bracteoles giving good cover. The scent of the cones can be described as fine, hoppy and a little spicy. The length of the vegetative period is the same as that of 'Lúčan', i.e. 122–127 days. Because 'Aromat' is resistant to *Peronospora,* this variety is recommended for use on sites where downy mildew is prevalent. This variety can be used anywhere in the Czech hop-growing regions.

The variety 'S i ř e m', consisting of 755 plants, was bred by a three-phase selection from the stock of the half-early red bine 'Žatecký krajový' growing in the community of Siřem. The above-ground part of the plant is cylindrical. It has stronger growth than the variety 'Aromat'. With narrow spacing the fertile shoots are produced at a height of 160–180 cm and at 220–225 cm with wide spacing. The bine has a green-red colour with big leaves. The stem is 10–12 mm in diameter and the distance between nodes is 27–28 cm. It climbs well. The shoots in the lower two thirds are 110–125 cm long and propendent, and in the upper third they are 55–90 cm long and semipropendent. The foliation of the shoots is moderate. The cones are egg-shaped, they have a prolate basal part, with a very regular structure and a good cover of bracts and bracteoles. The scent of the cones is correct, fine and hoppy. The length of the vegetative period is the same as that of 'Lúčan'. It is suitable for all Czechoslovak regions including warm sites and rich soils.

Four other recognized clones were bred by individual selection from the original regional variety 'Žatecký krajový'. A clone is the offspring of one original plant, propagated vegetatively. Hence its genetic base is very narrow so that the plants, and their final product – the cones – have, consequently a very high degree of equilibrium. Moreover the requirements of clones for agro-ecological conditions are very precise. In suitable conditions these clones have 15–20 per cent higher yields than the original variety 'Žatecký krajový'. If planted on very rich soils the cones may become overgrown.

'O s v a l d ' s   c l o n e   N o .   3 1 ' was bred in 1927 by individual from the population of the variety 'Žatecký krajový' in a hop garden in Rakovník, and was acknowledged in 1952. It is a half-early red bine with very tall growth. When grown with narrow spacing the fertile shoots occur at a height 140–180 cm, and, with wide spacing, at 220–240 cm. The bine is reddish in colour and has medium sized leaves. The shoots are mostly propendent, 90–110 cm long in the lower part of the bine, and 60–90 cm in the upper. The cones are egg-shaped, and have regular structure and cranking. The scent of cones is fine and hoppy. The length of the vegetative period is 3 days longer than that of 'Lúčan'. 'Osvald's clone No. 31' has particular requirements for soil moisture and is suitable for heavy soils and sites in the valleys of Czechoslovak hop-growing regions. Its resistance to Peronospora is about the same as that of 'Lúčan'.

'O s v a l d ' s   c l o n e   N o .   7 2 ' originated in the population of the variety 'Žatecký krajový' in a hop garden in Deštnice. Its selection began in 1927 and the clone was acknowledged in 1952. It is a half-early red bine with very vigorous growth. When grown at narrow spacing, the fertile shoots appear at a height 150–170 cm, and with wide spacing at 230–260 cm. The bine is reddish with a good foliage of big leaves. The shoots in the lower two thirds are 80–100 cm long and propendent and in upper third are semipropendent and 50–80 cm long. The cones are prolate egg-shaped and may even be four-sided with very rounded corners. Cone structure and cranking is very regular as is the covering of bracts and bracteoles. Its scent is correct, fine and hoppy. Because of their excellent quality the·cones of 'Osvald's clone No. 72' are used as the Czechoslovak and world standard of quality. The cones ripen 3 days earlier than those of 'Lúčan'. In confined sites and during rainy summers this variety is susceptible to attack by Peronospora. Its requirements for soil are not very demanding. This variety is suitable for open sites in Czechoslovak hop-growing regions. In wet soils and warmer regions it is disposed to overgrowing and proliferation of the cones.

'O s v a l d ' s   c l o n e   N o .   1 1 4 ' originated from the same stock in Deštnice and its development was similar to that of 'Osvald's clone No. 72'. It was acknowledged in 1952. It is a half-early red bine of medium height. When grown at narrow spacing the fertile shoots appear at a height of 130–150 cm, and with wide spacing at 200–240 cm. The bine has reddish colour with good foliage of medium-sized leaves. In the lower part of the plant the shoots are semipropendent and 70–100 cm long and in the upper third they are 50–70 cm long. The structure of cones and the cranking is regular. The cones are prolate especially those from hop-gardens in warm sites and humous soils. The scent is correct, fine and hoppy. 'Osvald's clone No. 114' ripens about 3 days earlier than the variety 'Lúčan'. It is suitable for planting in so-called field sites, and has proved excellent, especially in cretaceous marly soils in the Džbán terrace and in the Rakovník region. It is not suitable for very wet soils and sites, where its cones overgrow and proliferate and are endangered by *Peronospora*.

The variety 'Z l a t a n ' was bred by clone selection from the population of the variety 'Žatecký krajový' by the Research Institute of Hop Growing in Žatec as clone No. 118 and it was recognized in 1976. It is a half early red bine with very vigorous growth and cylindric habit of the above-ground parts. When grown at broad spacing its shoots occur at a height of 180–260 cm. The bine is of a light red-green colour. Its climbing is good, with an average of 3.8 windings per metre of length. The shoots are semipropendent. The cones are egg-shaped with a regular structure and a good covering of bracts and bracteoles. The scent is right, fine and hoppy. This variety is suitable for medium heavy and heavy soils with ample moisture. It is sensitive to overfertilizing with nitrogen. It causes a proliferation and overgrowth of cones.

ONTOGENESIS OF HOP PLANTS

A definition of ontogenesis, valid for all living beings, is that it is an evolutionary process concerning the total existence of the organism as an individual. It includes the complex of all structural and functional changes from its birth up to the decay of the individual organism.

Ontological changes in hop plants can be best observed on the perennial organs of the underground stem and on roots. Their annually repeated thickening leads to the production, each year, of one annual ring. Thus, according to the number of annual rings it is possible to determine their calendar (chronological) age. The mass of underground perennial organs increases with increasing plant age but the percentage of their viable tissues decreases after the fourth year. Therefore, the number of living dormant buds on healthy old wood does not increase after the fourth year. Later, the number of living dormant buds decreases especially if the old wood becomes diseased. Furthermore a decreasing number of one year old active roots and the thickness of the storage tissues in root tubers is important for the potential production of hop plants. Changes in these and other main functions, especially changes in the biological age of underground organs during ontogenesis, lead after the fourth year to a decrease in the potential fertility of the hop plant which is slow first but later becomes rapid. It is, however, possible to modify to a certain extent the consequences of biological ageing and to promote certain rejuvenating processes by suitable biological and agrotechnical interventions. The real fertility of hop plants can thus be maintained up to 15 years of age and that of the whole stock can be maintained at an economically acceptable level up to 20 or even 30 years, because plants are renewed and exchanged within the stock.

All this implies that the process of ontogenesis, including many different processes, especially biological ageing and rejuvenating, clearly affects the course of the annual life cycles of the individual hop plant. These cycles are, however, not the same, they change during ontogenesis and so they affect the annual economic results, i.e. the annual yield and quality of cones as well as the longer term stability of the yield and its quality. In order to become informed of the reasons for these changes, it is necessary to obtain knowledge of the theorical basis of ontogenesis and, as a result, to determine the criteria for its regulation.

The basis of ontogenesis is that of a successive differentiation of all of the material, formal and functional properties of the organism. These properties are obviously very numerous. Therefore it is necessary to define the main criteria for the evaluation of the whole of ontogenesis. Normally, with annual and biennial plants, the main criterion is the production of generative organs. It is not feasible to adopt this as the main criterion for hop, because the growth of generative organs and the general fertility of the hop plant is very variable and cannot be used as a reliable indicator of the whole ontogenetical process. Thus, for example, the yield of annual stocks from autumn seeding was four times higher than from spring seeding, according to investigations made by RYBÁČEK (1972). *The best of numerous ontological criteria were established as the chronological and biological age of the underground stem organs,* which are also the natural organs of vegetative propagation. This is usually established after the end of growth of the above-ground organs, i.e. after their total decay or after their decapitation for picking at harvest. At this time the individual hop plant is represented by its underground part, i.e. the underground system of stem and roots.

The life of an individual hop plant is affected, among other things, by the process of its origin and its decay. Every individual descends from a maternal organism regardless of whether it was produced by the vegetative or the generative type of propagation. In both cases it will become independent, either after the separation of a bud cutting from the mother plant or after the ripening of the fruit giving viable seed. The hop plant decays in two ways: in the first the individual decays totally, in the second it is the oldest underground stem organ, i.e. the original old wood, which decays. This is replaced by one or more underground stem organs, such as new wood, suckers or separate sectors of old wood.

The ontogenesis of the hop plant from its origin to its decay can be divided into individual periods. After long investigation, seven periods were established, according to the chronological and biological age of individual underground organs. These are shown in Fig. 39.

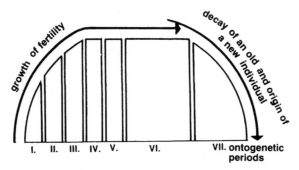

Fig. 39. Diagram of the ontogenic process.

Chronological and biological age of underground parts of the hop plant

The chronological age of the underground stem organs of the hop can be easily established as can that of any dicotyledonous plant, by a study of the secondary thickening of its stems and roots. Phloroglucinol is excellent as a stain for the annual rings, the number of which corresponds to the age, in years of that particular organ. The greatest age for a hop plant was 10 years, determined by NEČIPORČUK (1955) from the number of annual rings in the old wood. Investigations by RYBÁČEK (1967) on a large number of plants in an old hop garden revealed two plants with a provable age of 14 years.

The same method serves to establish the chronological age of roots, according to the number of annual rings near their point of eruption. The individual annual increments can be established from cross sections taken at short intervals along the root. It was discovered by RYBÁČEK (1967) that some roots did not elongate in certain years. This was due to the fact that their vessels had no direct connection with the above-ground system in these particular years and consequently the root had insufficient carbon nutrition.

Under the same growing conditions and in the same chronological year, the biological and agricultural properties of individual plants demonstrate considerable variation. Difference in their biological age is the deciding factor in these plants. The term "biological age" is more general and more precise than the term "physiological age", formerly used, because biological changes due to age affect not only physiological processes, but all biological and agricultural properties. According to RYBÁČEK (1966, 1975) biological age represents the counteracting processes of ageing and rejuvenation. The quantitative evaluation of these two processes can be derived from the ratio of nitrogen to calcium, because the changing content of calcium corresponds well with the process of growing old and the content of nitrogen, especially that in proteins, correlates, in a simplified way, with the processes of rejuvenation. Biological age (BA) is established according to the following formula:

$$BA = 10 \, Ca : N$$

Thus, it is possible to establish the biological age of individual tissues, organs, organ systems and the whole plant. The biological age of particular organs is measured by the ratio of old and young tissues, and that of an organ system depends on the ratio of young and old members of the system.

At harvest time RYBÁČEK (1968) found differences between the young wood (new wood and new suckers) and the mature wood (old wood and old suckers) of the underground stem systems of five-year old hop plants ('Osvald's clone No. 72'). These differences in biological age are shown in Table 6, which indicates that these are chronologically different groups of

TABLE 21a **Dry matter content in underground parts of the hop plant**

Plant age	Dry matter content (g)						
	old wood	old suckers	mature wood	new wood	new suckers	young wood	root-stock
1-year	8.59	–	8.59	4.78	–	4.78	13.37
2-years	22.00	0.70	22.70	12.56	3.80	16.36	39.06
3-years	39.84	2.44	42.27	14.91	5.81	15.42	57.69
4-years	46.84	2.66	49.50	18.14	8.93	27.06	76.56
5-years	51.15	13.36	64.51	18.38	14.47	32.85	97.36
6-years	72.64	9.21	81.85	14.87	10.89	25.76	107.61
14-years	142.14	5.09	147.23	22.97	45.74	68.71	215.94

TABLE 21b **Dry matter content in underground parts of the hop plant**

Plant age	Dry matter content (g)					
	tubers	horizontal roots	vertical roots	skeletal roots	total roots	total underground organs
1-year	–	2.43	13.68	16.09	16.09	29.46
2-years	0.22	38.55	7.88	46.46	46.68	85.71
3-years	6.81	53.56	14.98	68.54	73.35	133.04
4-years	8.24	67.62	19.05	86.77	94.91	171.47
5-years	16.65	70.27	19.10	89.37	106.03	203.39
6-years	32.22	73.43	22.92	96.34	128.56	236.17
14-years	74.77	222.06	15.88	237.93	312.70	528.64

TABLE 22a **Percentage of dry matter content in organs of underground parts of hop plants**

Plant age	Percentage of dry matter content					
	old wood	old suckers	mature wood	new wood	new suckers	young wood
1-year	29.16	–	29.16	16.23	–	16.23
2-years	25.67	0.80	26.47	14.65	4.43	19.08
3-years	29.95	1.83	31.78	11.22	4.37	15.59
4-years	27.32	1.55	28.87	10.58	5.21	15.79
5-years	25.15	6.57	31.72	9.04	7.11	16.15
6-years	30.75	3.90	34.65	6.30	4.61	10.91
14-years	26.89	0.96	27.85	4.35	8.65	13.00

TABLE 22b **Percentage of dry matter content in organs of underground parts of hop plants**

Plant age	Percentage of dry matter content					
	root-stook	tubers	horizontal roots	vertical roots	skeletal roots	total roots
1-year	45.39	–	8.20	46.44	54.61	54.61
2-years	45.55	0.26	44.98	9.19	54.17	54.45
3-years	47.37	5.12	40.26	11.26	51.52	52.63
4-years	44.66	4.81	39.44	11.11	50.55	55.34
5-years	47.87	8.19	34.55	9.39	43.94	52.13
6-years	45.56	13.64	31.09	9.70	40.79	54.44
14-years	40.85	14.14	42.01	3.00	45.01	59.15

organs (one-year old and five-year old). It is interesting to compare the organs of new wood and new suckers of the same chronological age shown in this Table. This clearly shows the lower biological age of the suckers.

It is similarly interesting to compare the biological age of the individual parts of the root system as shown in Table 8. The mass of dry matter and its distribution in both systems is shown in Tables 21 and 22.

Periods in the ontogenesis of the hop plant

The ontogenic categories introduced by RYBÁČEK (1968) are based on his investigations into the chronological and biological age of the underground stem system (rootstock) and of the oldest organs of hop plants, which is the wood, including its fundamental organs, the dormant buds. After a long investigation he introduced s e v e n   i n d e p e n d e n t   o n t o-g e n i c   p e r i o d s   o f   h o p   p l a n t, namely:
1. bud (embryonic) period,
2. strap-cutting and seedling period,
3. rooted-cutting period,
4. shoot period,
5. adolescence,
6. adulthood age and
7. old age.
The mass of the underground organs and the structure of underground parts of different ages of the hop plant are shown in Tables 21 and 22.

The bud (embryonic) period

This period starts at the origin of the zygote i.e. at the pollination of the female flower and lasts until the activation of the embryo at the time of germination of the one-seeded fruit. In the case of buds the period starts at the origination of the dormant bud and ends at the moment of bud activation.

The seedling and strap-cutting period

This period lasts from the start of germination of the seed, or from the moment of activation of a dormant bud to the formation of new wood with new buds. This period occurs during the first year of both generative and vegetative propagation and concerns not only seedlings but

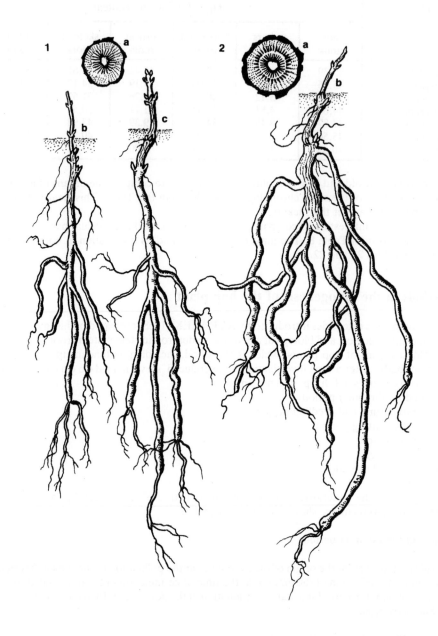

Fig. 40. Ontogenic periods of the hop plant: 1 – strap-cutting and seedling period, 2 – rooted-cutting period, 3 – adolescence period, 4 – adulthood period (a – cross section of rootstock, b – underground part of the plant, c – rooted-cutting).

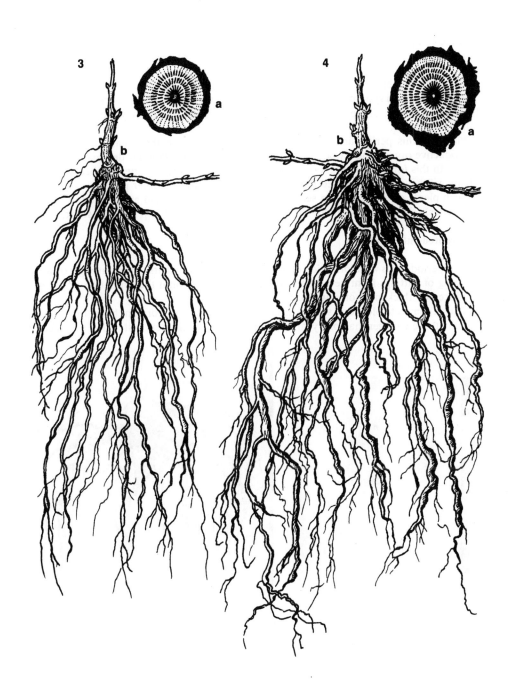

also rooted-cuttings derived from buds and underground growth. In both cases the underground parts become modified during their first growing season and new wood is produced on which a few dormant buds, younger than one year, are developed. The rootstock does not become completely developed during this period, because not all of its components such as dormant buds, old wood and new and old suckers are able to be produced. The only complete part is the root system which, however is not very extensive. A strong apical dominance develops during this period and therefore the above-ground system consists of only one bine. The above-ground system is usually incomplete, having no reproductive organs, sometimes it has no side shoots. The plants are conseguently either infertile or they produce only a small number of cones.

Plants of this period are designated as seedlings because their main maintenance and reproductive organ is new wood, which when used for propagation is designated as seedling wood. In this form of propagation, ontogenesis begins in the next period, because the new plants will have already passed the seedling period on their mother plants.

Such plants developing from seeds are known as *seedlings* and are planted in seed plots (nurseries) to provide shoots for the intensive propagation of hop planting stock.

The rooted-cutting period

This period starts with formation of new wood on mother plants (strap-cuttings) or with the planting of cuttings from new wood or suckers and lasts until the formation of old wood, produced by thickening of new wood.

Thus a complete old wood system has two age categories of dormant buds, those one year old and those younger than one year. This time underground stem system has no suckers but

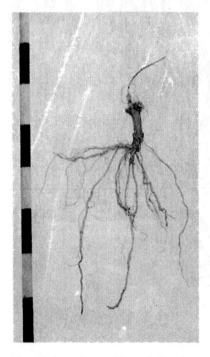

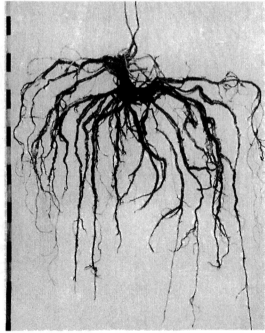

Fig. 41. The shape of hop during the rooted-cutting period.

Fig. 42. Underground system of a four-year old hop plant.

84

their buds are already formed on the old wood. When seedlings develop shoots a complete above-ground system is likely to be formed, with a large number of bines. Plants which originate from suckers show strong apical dominance and produce an one-bine, above-ground system which becomes complete if the growing conditions are good, but otherwise it will not form reproductive organs (inflorescence and syncarpy). During the pot period there is an increase in the total number of roots including tap roots.

The plants of this period are usually designated as one-year, pot or virgin plants. When they are fertile, their yield is very low and they produce only so-called virgin cones. They are planted in pot plots as *rooted-cuttings derived from new wood or suckers.*

The period of shoots

During this period a third age group of dormant buds appears on the old wood and at the same time the suckers begin to develop from their buds. The underground stem system (rootstock) however is not yet complete, since it lacks one annual set of dormant buds and has no old suckers. The complete above-ground system consists of a large number of bines which, at this stage have not yet reached full size and therefore the yield of cones is relatively low. The plants usually grow to reach the top of the stringing structure.

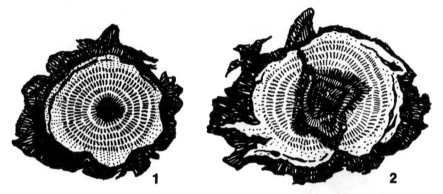

Fig. 43. Old wood at the sectoring: 1 – beginning of sector formation, 2 – two sectors completed.

The period of adolescence

This period features the completion of the underground stem system. A fourth cycle of underground dormant buds appears and the new suckers become old. At the end of this period the rootstock system is complete and possesses all requisite organs. Healthy well-grown and well developed plants have a robust underground system which can suport a full yield of cones either in this period or at least at the beginning of the next.

The period of maturity

This period is in the fourth year of the plant age and featured by old wood and with a stable number of age groups of dormants buds. The older dormant buds on the old wood decay and in their fifth year are replaced by new buds. Healthy and well-grown plants maintain a relatively stable sized above-ground system with a number of bines and similar of cones. Certain changes occur however in the underground stem system, where there is an increasing rate of tissue death in the old wood and old suckers. Even greater changes occur in the root system, where not only branch roots but also tap roots decay and are replaced. The dead internal tissues in the old wood collapse leaving cavities. This procedure represents the end of the maturity period and the beginning of the senility period.

The period of senility

This is the period from the start of collapse of internal tissues in the old wood up to the total decay of the plant. Natural decay of an individual hop plant occurs if the old wood splits to provide two or more independent tap roots. In this case the former single individual becomes two or more. The above-ground system of a senile plants is reduced and the cone yield decreases before the plant finally collapses.

Progressive and ontogenic heterogeneity

The arrival of new individuals in a homogeneous stock disturbs its homogeneity. The whole of a hop stock is *homogeneous with regard to age* after the plantig of seedlings or rooted-cuttings, but its *heterogeneity* progressively increases from the second year onwards. The reason to this is that particular plants fail to grow normally and others decay immediately after planting or later. Their replacement by younger plants improves hop stock and represents an *artificial rejuvenation* but at the same time it causes genetic heterogeneity. However rejuvenation is not only artifical, as RYBÁČEK (1967) discovered, and hop plants are capable of *spontaneous rejuvenation*. A new base arises when wood grows from a bud on the old decaying wood and eventually serves as its replacement. Thus there is restitution of the old wood as well as the tap root.

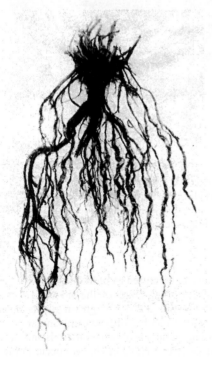

Fig. 44. Underground part of fourteen year old plants.

Another type of spontaneous rejuvenation was discovered and described by NEČIPOR-ČUK (1955). This involves the sectoring of old wood. Such sectors come to represent independent individuals of a slightly lower age than the original plants.

The rejuvenation of stock in these various ways causes diversity in growth and leads to ontogenetic heterogeneity in the hop stock, but this has a positive effect in prolonging the life of the stock. On the other hand, however, such ontogenetic diversity is disadvantageous, because it lowers the total number of efficient well-grown plants.

If a mature stock is homogeneous, then all plants are in the adult age but heterogeneous stock will include rooted-cuttings, adolescent and senile plants each of which has a potential yield. Such a heterogeneous stock may be composed of up to three "generations" of vegetatively propagated plants. Therefore, the average age of the plants in the stock is lowered with regard to the calendar age of such a hop garden. In a fourteen-year old hop garden in Ruda RYBÁČEK (1967) found the average age of plants to be 10.07 years with the most numerous being 10-11 and 12-year old plants, though there were 14 year old plants. In the 36-year old hop garden in Vědomice, the average age of the plants was 8.06 year. Plants 8 and 9-year old were most numerous, but the oldest were 11-year old. RYBÁČEK also found that, as the hop garden grows, there is an increasing danger that the central part of wood will decay, that the ontogenesis of the stock will be completed and the hop plants will die in increasing numbers.

The greater the age of the hop garden where the old wood of the individual plants has decayed the wider is the spacing within the rows. This wide spacing also appears in adolescent hop stock, but there it has a different cause, mostly resulting from poor rooting or from the destruction of young adolescent plants.

Regulated ontogenesis of hop plants

It is mostly the size and structure of the underground system which is regulated during the ontogenesis of the hop plant. The end result of this regulation is apparent in the quality and quantity of cones during each annual cycle throughout the whole ontogenesis of the perennial hop plants. Therefore it is necesary to understand the phases of regulation in adolescent, adult and senile plants.

In *the ontogenetically non-adult plants,* i.e. during the period of seeding, rooted cutting or adolescence, the regulation is mainly focused on improving the size of the root system. This in turn is dependent on the size of the above-ground system and particularly on the amount and duration of photosynthetic activity and its apparatus, i.e. especially on the leaves. The experiments by RYBÁČEK (1978) showed that the function of leaves can be increased, photosynthesis is more efficient and the migration of substances to the underground system is greater if flowering of the hop plant is impeded. Flowering and cone production is undesirable during the seedling and root-cutting periods. We do not reduce the rootstock during the adolescent period and we do not hinder the natural decay of the above-ground system in any period so as to accumulate maximum reserves of materials in underground systems.

The severing of the top growth of *the adult hop plant* at harvest raises the biological age because the limited migration of substances limits the growth of new tissues and the rejuvenation of the old. In the rootstock, the situation is worse because the annual cutting removes the new young parts (new wood and new suckers) and the rootstock is then practically reduced to the old wood.

The only real solution is to improve the rejuvenation of the underground system. It is possible, however, to shift the harvest time towards the beginning of the period of full technical maturity and thus to limit negative consequences of harvest by decapitation. The problem can also be lessened by leaving as much bine as possible after harvest to help with the rejuvenation of the root system. The regeneration of roots also can be improved by inter-row hoeing followed by the cutting of horizontal roots.

ANNUAL CYCLE OF THE HOP PLANT

The ontogenetic development of the hop plant spreads over many years and includes repeated annual cycles. The beginning of an annual cycle can be taken as the termination of material transport from the above-ground parts of the plant. With the total removal of top growth by decapitation at harvest, the material migration is abruptly stopped, so the beginning of a new annual cycle starts, precisely at the time of harvest.

TABLE 23a   **Biological age (index 10 Ca:N) of above-ground parts of three-year old hop plants measured fortnightly in 1972**

Plant part	Date			
	29. 5.	12. 6.	26. 6.	10. 7.
Bines	3.57	5.24	7.87	11.64
Shoots			4.92	9.55
Bines + shoots	3.57	5.24	6.90	11.12
Leaves on bines			7.21	9.73
Leaves on shoots				5.58
Total leaves			7.21	8.28
Cones				
Cones + Leaves			7.21	8.28
Above-ground part	3.57	5.24	7.08	9.57

TABLE 23b   **Biological age (index 10 Ca:N) of above-ground parts of three-year hop plants fortnightly measured (Means of results in 1972 and 1973)**

Plant part	Date			
	24. 7.	7. 8.	21. 8.	4. 9.
Bines	15.59	19.66	21.81	23.20
Shoots	13.17	12.46	15.61	21.81
Bines + Shoots	14.74	17.07	18.85	22.73
Leaves on bines	14.86	13.69	16.32	20.61
Leaves on shoots	8.56	10.49	14.11	16.96
Total leaves	12.45	12.05	15.14	18.60
Cones	4.01	3.96	4.24	4.76
Cones + leaves	10.04	8.20	10.14	11.90
Above-ground part	11.91	11.50	13.26	16.29

Basic processes during the annual cycle

During the annual cycle as in the whole process of ontogenesis, the material, morphological and functional properties of the whole plant become gradually differentiated. This differentiation is most obvious on the above-ground part of the plant which comes into existence and decays in an annual cycle. This cycle is also accompanied by changes in both underground systems. Every differentiation is preceded by certain preparatory processes which can be detected in the preceding vegetation, from whence the products of photosynthesis migrate to provide the basis of changes in the underground system.

The annual cycle of the hop plant has many parts therefore it is necessary to pay particular attention to its main features. These mainly involve metabolic procedures, the accumulation and reduction of dry matter (production), propagation (reproduction) and the changes associated with ageing. There is no space in this publication for the detailed analysis of physiological processes therefore attention will be given to certain special features of the hop plant.

Changes in biological age during the annual cycle

The biological age changes rapidly during growing season but also less quickly at other times. The changes do not occur at the same speed in all above-ground organs as can be seen from Table 23. The most rapid changes occur in the leaves of shoots, followed by the shoots themselves, and changes are slower in leaves on bines and in the bines themselves. The Table also shows the beginning of the generative period (from 10th July in Czechoslovakia) clearly increased the biological age of all of the vegetative organs present at that time.

Special properties of photosynthesis

The basic metabolic process in hop plants, as in other green plants, is *carbon metabolism,* with its two components, *photosyhthesis* and *respiration.* When the carbon metabolism has a positive balance there is an increase in dry matter. The process of photosynthesis is said to have three basic components: *a. P. t.,*
where

a = intensity of photosynthesis;
P = surface of photosynthetic organs;
t = time interval of photosynthetic process.

The whole above-ground system of the hop plant is adapted for intensive photosynthesis, because all above-ground organs (except of fruits) are modified for this process. Main organs of photosynthesis are, however, the leaves on bines and shoots. The productivity of photosynthesis represents the balance between photosynthesis and respiration and is usually designated as a net photosynthetic efficiency *(NAR).*

Normally, leaves with high photosynthetic activity have many stomata, a dense nervature and a palisade parenchyma with several strata. Hop leaves have two of these properties: they have a density of stomata per square centimetre of leaf surface approximately four times greater than that of sugar beet, and the branching of the vascular system is very dense. The palisade parenchyma however, has only one layer. The quality of plastid colour is especially important with regard to the internal properties of leaves. They change with age so that the biological age *(BA)* of leaves can be estimated by using the changes as an indicator.

A quantitative indicator of foliar activity is the surface area (cm^2) of leaves of bines and shoots growing on one bine *(A).* Other indicators such as the foliar surface per m^2 of the stock *(LAI),* are derived from this value. The relative efficiency of the photosynthetic apparatus *(LAR)* is usually determined as the ratio between dry matter *(W)* and foliar surface *(A).*

These values are shown in Table 24 from which it is evident that with increasing biological age of leaves the net photosynthetic efficiency *(NAR)* decreases.

TABLE 24   **Growth analysis of above-ground parts on three-year old plants (Means of results from 1972 and 1973)**

Date	Wl/W	A/Wl	LAR	R	R_A	NAR
29. 5.	0.53	2.19	1.16	–	–	–
12. 6.	0.57	1.79	1.03	0.086	0.077	0.079
26. 6.	0.64	1.67	1.06	0.074	0.072	0.068
10. 7.	0.56	1.34	0.75	0.046	0.021	0.053
24.7	0.52	1.93	0.79	0.022	0.031	0.029
7. 8.	0.42	2.37	0.77	0.008	0.006	0.010
21. 8.	0.34	2.82	0.96	0.013	0.028	0.014
4. 9.	0.33	3.46	1.13	0.005	0.017	0.005

Wl = dry matter of leaves (g)
W = dry matter of above-ground plant parts (g)
R = specific speed of growth of above-ground plant parts
R_A = specific speed of growth of leaf area
A = assimilation surface (cm²)
LAR = leaf area ratio
NAR = net assimilation rate

Production processes

With a positive balance between photosynthetic and respiratory processes the mass of dry matter increases as does the amount of living material the so-called biomass – and these, together are designated as the biological production of the whole plant. With a negative balance the dry matter and biomass decrease and the biological production is lower.

During the whole of the growing season, from the appearance of the above-ground shoots to the time of their decay in the autumn, day time increments alternate with night time decrements. During the rest of the annual cycle, there are decrements not only during the night but also during the day as a result of respiration. As well as the total balance based on the dry matter of the whole plant, it is possible to follow the increments (accumulation) or decrements (reduction) of dry matter in individual organs and their respective systems.

In terms of plant production, *the increments (accumulation) of dry matter in the whole plant* is the most important feature of the photosynthetical and respiratory balance *(NAR = 0).* This accumulation is possible because under favourable conditions the intensity of photosynthesis exceeds that of respiration by 20–30 times. The changes in mass and distribution of dry matter are shown in Table 25.

This Table makes it clear that the bine leaves are the first above-ground organs to finish their growth, and this is at the same time as the start of growth of the generative organs (inflorescence and cones, 10th July).

The total mass of both systems of underground organs remained at about the same level up to 7th August and then rapidly increased mainly due to the growth of young wood and roots.

Dry matter increments are reflected in the lengthening and thickening of individual organs, both of which are measurable. The growth of the bine, which is the first above-ground organ to start to grow is measured in terms of height increments. This growth measurement is reliable only until the shoots begin vigorous growth, because then the elongation involves shoots as well as bines. The thickening process is derived from measurements of the diameter of organs, always measured at the same place. The decrement of dry matter can only be determined by weighings.

Decrements in the dry matter of whole plants are due to respiration in all living parts of the

TABLE 25a Average amount of dry matter obtained from three-year old plants (Means of results from 1972 and 1973)

Plant part	Amount of dry matter (grams) to date			
	29. 5.	12. 6.	26. 6.	10. 7.
Bines	11.34	30.39	59.22	138.22
Shoots	0.11	3.76	19.72	51.06
Bines + shoots	11.45	34.15	78.94	189.28
Leaves on bines	12.78	43.37	116.51	153.48
Leaves on shoots	–	2.66	23.42	70.37
Leaves on bines + leaves on shoots	12.78	46.03	139.93	223.85
Cones	–	–	–	6.17
Cones + leaves	12.78	46.03	139.93	230.02
Total above – ground part	24.23	80.17	218.76	419.29

TABLE 25b Average amount of dry matter obtained from three-year old plants (Means of results from 1972 and 1973)

Plant part	Amount of dry matter (grams) to date			
	24. 7.	7. 8.	21. 8.	4. 9.
Bines	170.96	166.71	197.65	200.16
Shoots	85.35	90.15	95.54	125.64
Bines + shoots	256.30	256.85	293.19	325.80
Leaves on bines	155.99	114.47	137.03	131.51
Leaves on shoots	97.69	104.24	136.31	149.43
Leaves on bines + leaves on shoots	248.68	218.71	273.34	280.94
Cones	101.68	199.75	235.57	255.96
Cones + leaves	350.36	418.46	508.91	536.89
Total above – ground part	606.66	675.31	802.09	862.69

TABLE 25c Average amount of dry matter obtained from three-year old plants (Means of results from 1972 and 1973)

Plant part	Amount of dry matter (grams) to date			
	29. 5.	12. 6.	26. 6.	10. 7.
Total above – ground part	24.23	80.17	218.86	419.29
Old wood	31.38	27.85	25.58	24.23
Young wood	1.56	2.90	6.26	2.25
Old + young wood	32.94	30.75	31.84	26.48
Roots	63.86	60.16	62.89	63.96
Total underground part	96.80	90.91	94.72	90.43
Total above + under ground parts	121.03	171.08	313.58	509.72

Average amount of dry matter obtained from three-year old plants (Means of results from 1972 and 1973)

Plant part	Amount of dry matter (grams) to date			
	24. 7.	7. 8.	21. 8.	4. 9.
Total above – ground part	606.66	675.31	802.09	862.69
Old wood	25.23	21.28	27.47	29.11
Young wood	4.37	8.18	16.39	18.55
Old + young wood	29.60	29.46	43.86	47.66
Roots	68.25	58.62	80.62	141.36
Total underground part	97.84	88.08	124.48	189.01
Total above + underground parts	704.49	763.39	926.53	1 051.70

plants and these decrements are homogeneous. Intermittent decrements, on the other hand are due to the death and shedding of whole organs and their systems. The shedding involves mainly the leaves and the terminal rootlets die. Considerable decrements of dry matter in particular organs result from material transport (redistribution). This material redistribution towards the end of the vegetative period is very important because nitrogenous, nitrogen-free and certain mineral materials migrate from above-ground to underground organs. A reverse migration of these groups occurs at the beginning of the vegetative period. As well as these major redistributions between above-ground and underground organs, certain minor redistributions occur in the underground and among leaves during the vegetative period (PICHL, RYBÁČEK, 1972).

Reproduction

The result of growth processes is the vegetative biomass, otherwise known as biological production or the biological crop. Part of this vegetative biomass can, under favourable conditions, be used to create new individuals by propagation. *Vegetative propagation* is a simple manner of reproduction very common in the plant world, *Generative propagation* is, biologically a more sophisticated method of reproduction. The hop plant uses both means of reproduction and both are economically important: generative reproduction for the main product – hop cones, and vegetative reproduction providing stock and for improving old ones. Special organs of vegetative propagation, formed in the hop as underground runners (stolons) are known as suckers. Their origin and development is closer to the normal processes of growth, from a biological point of view, than is the formation of generative organs of propagation. The generation and formation of these organs demand favourable external and internal conditions which are different from the optimum conditions for vegetative growth.

Growth and reproductive development of hop plants

Vegetative growth and reproductive development are vitally important processes for the formation of above-ground and underground parts of the hop plant, both from the biological and the industrial point of view.

According to DOSTÁL (1962) growth is the vital function for the productivity of plants.

According to him, growth is a quantitative as well as a qualitative process which naturally modifies plants in close accord with the conditions of the internal and external environments.

Natural organs of vegetative propagation are all organs of the underground stem system. Every year new dormant buds are generated in this system as is the new wood which grows vertically from the activated buds of bines, and as are the new suckers which grow horizontally from the buds of the suckers. These natural organs of vegetative propagation begin to develop earlier than the generative organs of propagation. The formation of seedlings and suckers is affected by agrotechnical interventions. The number of pieces of new wood is given by the number of development of bines, and the best indicator of the size of bines is their thickness, which is usually measured at the central point of the topmost internodes. The thickness of suckers is normally determined at the center of their total length. The number of suckers is affected by the total size of the old wood and by the number of old suckers remaining on the rootstock after cutting.

The size of increments of generative organs of propagation can be established from data obtained by the analysis of growth. It has been shown that the substances important for the formation of reproductive organs are produced in the green, above-ground organs i.e. mainly in leaves but also partially in the stems. On the basis of this RYBÁČEK (1975) elaborated methods for the analysis of reproduction.

The coefficient of reproductive processes is calculated according to the general equation:

$$KRP = \frac{W_{R^2} - W_{R^1}}{t_2 - t_1} - \frac{100}{WL + s},$$

where: $W_{R^2} - W_{R^1}$ = increment of dry matter in reproductive organs (grams);
$t_2 - t_1$ = time interval between two samplings of dry matter (days);
$WL + s$ = average dry matter increase (in grams) in leaves and stems between two samplings.

For the analysis of vegetative reproductive organs (calculation of *KRP*veg) it is possible to use the results of a reduced analysis of growth in the above equation. In this case dry matter for the above-ground organs as well as new wood and suckers is determined. This reduced analysis of growth is advantageous. because it is non-destructive of plants in the stock. This general equation can be also used for generative reproductive analysis (*KRP*gen) when the organs R_1 nad R_2 will be the cones.

A simultaneous study of the reproductive processes and the establishment of *KRP*veg and *KRP*gen makes it possible to compare their dynamics.

Generative reproduction involves considerable changes in the function and structure of vegetative organs. Indeed a whole new system of above-ground generative organs is formed. In the hop this system comes into existence quite suddenly and its formation depends on certain conditions. These conditions are numerous and it is not easy to select those which are most important for the change from vegetative to generative period, because vegetative propagation makes it possible for hop plants to survive for many years without generative propagation.

This transient period depends on the balance between internal and external conditions and only their mutual complementation makes the change from the vegetative to the generative period possible. If this complementarity does not exist, then fertile and infertile plants will be found in the same stock. This is particulary true of stocks of immature plants from the seedling period up to adolescence. It has been found, however, that seedlings as well as rooted-cuttings, can be fertile as early as their first vegetative period of ontogenesis. The first condition for fertility is a certain maturity of the plant, and according to RYBÁČEK (1967) this condition in rooted cuttings is a minimum surface of leaves on bine exceeding 1000 cm². This leaf area on bines cannot be substituted by the same or a larger leaf surface on shoots produced after the interruption of apical dominance by the removal of the tip of sprouted bines. Well grown hop plants normaly have a leaf surface on bines much greater than 1000 cm² before the generative period i.e. at the time when flower buds (catkins) appear. Therefore this information concerning the minimum surface of leaves on bines cannot be regarded as generally valid for all ontogenetic period.

Among external factors, the intensity of illumination has a powerful influence. MAR-KENSCHLANGER (1934) regarded full illumination as the main condition for the hop to pass to the fruiting phase (generative period). A minimum light intensity is necessary to stop the shedding of buds and inflorescence and to produce cones. Higher light intensity increases the number of cones, it reduces the plant height at which they will be developed and it accelerates the ripening process.

Investigations concerning the effect of daylight have shown that a lengthening as well as a shortening of the daylight period starts the generative process.

All interventions which accelerate the vegetation and the biological senescence of the above-ground system of hop plants cause an earlier commencement of the generative period. All interventions which retard vegetation, and which slow down the biological senescence of above-ground organs, increase the density of setting of cones. If this retardation is so great that the migration of storage materials to the underground organs occurs before the generative period, then no generative organs appear and such bines remain infertile. This situation can be found on a retarded or shaded bine during any ontogenetical period of the hop plant. This demonstrates priority given to self-preservation and to the vegetative propagation of perennial hop plants over generative propagation. According to the results of experiments done by RYBÁČIK (1967) the migration of materials to the underground organs is significantly affected by short days. Long days or uninterrupted light hinders this migration and not only starts generative period but also guarantees more cones.

Periods of the annual cycle of the hop plant

Large changes due to the annual cycle are particularly evident on underground organs e.g. on activated buds and also on all these above-ground organs which decay every year. The annual cycle of the hop plant becomes most evident in the different levels of intensity at which the individual organs grow. Accordingly t w o   d i f f e r e n t   p e r i o d s   c a n   b e   d i s t i n-g u i s h e d :
1. period of cryptovegetation (dormancy) and
2. period of vegetation.

The period of cryptovegetation (dormancy)

This period lasts for approximately six months, from the second half of October to the beginning of April. It can be exactly measured from decay of the above-ground organs in autumn to the sprouting of new shoots in spring. The whole period can be divided into four phases (see Fig. 45).

a) *Preparatory period (predormancy)*. This period starts in the second half of October with the decay of the above-ground bine. At this time all buds on the rootstock, specially the activated buds on the new wood, interrupt their growth and enter the dormancy stage. The epidermis of the upper bracteoles and buds thickens and so the vegetative tip within the buds is better protected against winter weather. At the same time the non-atrophied ends of rootlets, which are white during the growing season, prepare for winter, and brown protection layers appear on their surface.

b) *Period of deep rest (dormancy)*. During hibernation buds on the root-stock cannot be encouraged into visible growth by any improvement of external conditions (e.g. in a glasshouse). No changes in growth can be detected and, consequently no morphological changes can be found in the underground parts of the plant. The processing of organic materials ceases, all life processes are suppressed to the highest possible degree, but they do not entirely stop. All living underground organs preserve their viability which will become apparent later. This period usually ends in December.

c) *Period of enforced rest (postdormancy)*. This period comes after the period of deep rest from which it is distinguished in that the growth of buds on rootstock is no longer hindered by the internal properties of the plant but by unfavourable external conditions, usually low temperature. Despite the fact that the growth of buds cannot be observed and measured, certain important processes occur in the rootstock. These involve the transformation of reserve stock (e.g. polysaccharides are changed into monosaccharides), which thus become mobile and are transferred from the root tubers and roots to the rootstock. Also at this time the absorptive activity of the roots becomes reestablished.

d) *Period of underground germination and growth*. This period starts with the activation of buds on the rootstock and ends with the appearance of sprouts above the ground. During this period underground growth, in the darkness, is at the cost of reserve stocks accumulated during the previous year. The greatest speed of growth is in those shoots from the buds at the tips of new wood and suckers. These tissues are not far from the soil surface and are exposed to sudden changes of temperature and humidity. Particular meristems produce, underground, five to six internodes. The period of underground growth differs according to the temperatu-

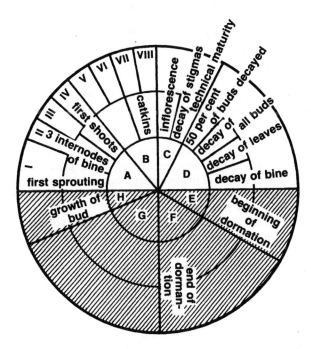

Fig. 45. Diagram of the annual cycle of the hop.

re, soil properties and thickness of the soil layer above the rootstock. Low temperature and a thick layer of heavy soil above the rootstock naturally lengthens the period of underground growth and may, consequently, lengthen the whole period of winter rest of the hop plant.

Fig. 46. Initiation and sprouting of hop bine: 1 – activated bud, 2 – proliferating bud, 3 – underground growth, 4 – sprouted shoot.

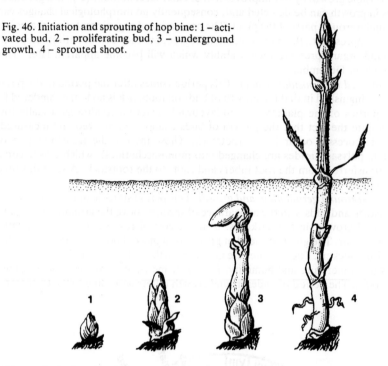

The period of vegetation

This period starts in spring, with sprouting of the bines, and lasts until their total decay in the autumn. The whole can be divided into eight separate periods according to the phases of plant growth.

The sprouting of the hop plant is a most important time because it is then that photosynthesis begins in green bines and leaves. The original source of organic materials in the storage organs is gradually complemented by the organic compounds produced by photosynthesis. This new source of organic matter becomes so prevalent that it can provide the materials for growth, regeneration of tissues, transport of solutions, respiration and even for the accumulation of new reserves.

a) *The period of linear growth of bines.* This period starts at sprouting and finishes with completed growth of three above-ground internodes of normal length. During this period the bines grow linearly. At the beginning they grow straight upwards then later, they begin to stoop. They are brittle, with a high water content and can be very easily broken. All three of internodes below the grown top are capable of producing new terminal growth. Each internode is approximately 20 cm long so that during this period the bines reach a total length of 60 cm (see Fig. 46. and 47). The activity of the root system becomes increasingly intensive during this period and the root-hairs begin to take up nutrients. Likewise, photosynthesis increases slightly, though the reserves continue to serve as the main source of organic substances.

b) *The period of establishment.* During this period the bines start to show their ability to climb around the support. This period begins once three internodes have been formed and lasts until beginning of shoot formation on bines, during which time the bines grow by 1 m (from 60 cm to 160 cm). After the three-internodes phase hop bines begin to grow in such a way that the tip of the bine moves anticlockwise in space, circumscribing circles of 10 to 15 cm diameter until it finds a support. One turn of the spiral takes approximately one hour.

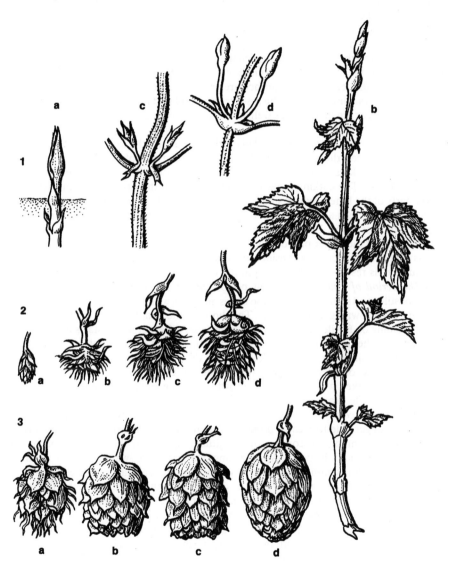

Fig. 47. Phenological phases of the hop plant: 1 – growth of the above ground plant parts (a – sprouting, b – growth of three internodes of the bine, c – start of shoot growth, d – start of catkin growth), 2 – development of inflorescence (a – 1/4, b – 1/2, c – 3/4, d – fully grown inflorescence), 3 – development of cone (a – 1/4, b – 1/2, c – 3/4, d – fully developed cone).

When the support is encountered the speed of winding decreases and its diameter becomes that of the thickness of the support. Bines on supports grow more rapidly than those without support, which creep and trail on the ground. In favourable temperatures bines grow very quickly and this can make their training in hop gardens very difficult. At this time the root system enters a period of intensive growth and the bases of horizontal summer roots originate in the underground parts of bines and suckers.

c) *The period of shoots and shooting.* During this period there is intensive growth of shoots. It starts at the time of first appearance of shoots and lasts to the time when the bases of inflorescences become visible. These bases are the so-called catkins. Bines grow very intensively during this period. Under favourable temperature conditions (24 °C), KLAPAL (1928) noted increments of up to 36 cm per 24 h. The shoots appear in the lower part of the bine after the growth of the main leaves on the bine. In the upper part of the bine, at a height of more than 3 m, the shoots appear simultaneously with the opening of the leaves and grow concurrently with them. The leaf surface area rapidly increases and photosynthetic productivity also increases. Nevertheless, the reserves accumulated in the underground organs during the preceding year are involved, as well as the shoots in the early stages of rapid growth of the bine.

d) *The catkins stage (butonization).* This period starts with the formation of the visible bases of the inflorescences, the so-called catkins and ends at the start of inflorescence formation. On male plants the catkins appear individually or in small groups in the axils of leaves at the nodes of fertile shoots. The catkins grow simultaneously with the branches of the fertile shoots. Lower parts of the bine produce smaller numbers of fertile shoots, whilst fertile branches do not branch very much and they produce small numbers of mainly individual catkins on their younger apical sections. Fertile branches, with groups of catkins in the axiles of their leaves are produced on the upper part of bine at every node of the shoot. Photosynthetic activity as well as root activity increases considerably during this period and the level of organic matter accumulated as reserves in the underground organs rises slowly.

e) *The period of inflorescence.* This period can be said to begin with the appearance of groups of stiles and stigmas in small hop cones and ends with their decay as is evident from the withering and browning of the stigmas. The shoots in the middle part of the bine blossom first, but blossoming on fertile branches which grow from different nodes starts at different times. Thus, the branches growing from the second and third node of the shoots i.e. from the axils of the second and third pairs of leaves blossom first, the others flower later. The rate of growth in length decreases considerably during this period but the growth of shoots and branching is very intensive, linked with an intensive growth of young leaves and an intensive growth and branching of fertile branches.

f) *The period of cone formation.* This period starts with the withering of the stigmas, when they are no longer able to receive pollen and it ends at the time of technical maturity of the crop. At the beginning the cones grow longitudinally, but at the end they become tightly closed, colourful and heavier. Approaching maturity, the water content of the cones decreases, and this is significant for brewering as the bitter substances and essential oils reach their optimal quality and quantity and the content of tannins remains high. The substances required by the brewer reach their optimum state at the time of technical maturity. During this period the growth of bines and shoots ceases and the growth of the youngest assimilative leaves is slower. But the cones continue to grow more vigorously, and at the same time the deposition of reserves in the underground organs increases, reaching approximately half of the final deposited amount. In the root system, the weight of root tubers increases, but the activity of roots declines and the number of fine rootlets is reduced.

g) *The period of physiological ripening.* In this period the cones ripen, as to the fruits i.e. achenes (stones) they contain. This period starts at the time of technical maturity of the cones and ends with the phase of physiologically complete maturity of the seeds. The cones become brown, the bracts turn outwards and are easily dislodged, and the quality as well as the quantity of main substances decreases. The growth of all above-ground organs stops and, beginning with leaves and shoots, the decay of particular organs starts.

h) *The period of decaying bines.* During this period, all of the above-ground organs perish. This period starts with the phase of physiological ripening of the cones and ends with complete decay of bines. This process of decay starts with the top of bine and the upper shoots, it continues downwards finally reaching the part of the bine near to the ground. The leaves on this lower part of the bine had ceased to function during the vegetative period but dormant buds had remained in their axils and these are the last to decay with the lower parts of the bine. Coincident with the decay of the bines, the transfer of reserves to the underground organs ends.

Heterogeneous age and reproductivity of above-ground organs

In the spring those plants not divided by cutting produce a large number of shoots. According to RYBÁČEK (1962) the number of shoots in the variety 'Žatecký krajový' in 1957 reached an average 27.6 per plant. The height of shoots before tying-in ranged from 0.69 to 1.0 m, and there was a difference in age of the individual shoots. A spring cut (made 12th April) reduced the number of sprouted bines to 9.4 per plant and improved their homogeneity. The number of bines was reduced to 4 by sawing and their homogeneity was improved by suitable selection. The pair of bines tied-in to the common string must be of the same length and age. Another pair of shorter bines but of the same length and age can be tied-in to the next string. This is important, because the shorter bine would be shaded, and therefore weakened, if a pair of bines of different length was set up to the same string. Such a pair of bines would flower unevenly and consequently the age and reproductivity of the bines would be more heterogeneous.

A certain natural reproductive heterogeneity appears on individual bines after flowering has begun. The shoots in the second third of the bine flower first and those in the lower and upper parts start 3 to 5 days later. The flowering of the shoots in the lower part can also be delayed because of shading. This delay then continues to the time of ripening when that of the two upper thirds will be in balance but the lower third have become further retarded because they have been shaded throughout the season. There is greater heterogeneity in those hop bines which have a second flowering one to two weeks later than the first.

Controlling the course of the annual cycle

The maximum yield of cones requires an optimum balance between the vegetative system and generative reproduction. This balance is reflected in the *coefficient of the reproductive process (KRP)*. There is a correlation between this coefficient and the total size of the above-ground part, so that there is a low *KRP* for plants with very large or very small above-ground parts. Therefore, if the total size of above-ground part is kept under control, then so is its coefficient *KRP*. The size of the above-ground part is affected by the quantity of reserves in the underground organs but more significantly by nutrition, by temperature and by watering. The amount of reserve substance in the underground organs depends on the date and height at which the plant is decapitated as well as on the extent of material migration from what remains of the bines. RYBÁČEK (1960) showed that well grown hop plants can accumulate in their underground organs bigger reserves than are needed for the following year providing that the normal processes of decay and material migration are not disturbed. Decapitation lowers the level of reserves to approximately fifty per cent if decapitation for harvest coincides with the time of full technical maturity, but more if decapitation is earlier. Therefore a higher decapitation is desirable so as to leave the lower parts of bines and shoots intact up to the time of total decay and thus augments the volume of reserves translocated to the underground organs. If it should be found that the reserves are insufficient, it may be necessary to support the growth of leaves at the beginning of the season by high dose of nitrogenous fertilizers, which will then promote the productivity of photosynthesis in the following period.

Of the total of sprouted bines only four are utilized and tied-in and therefore, at the start of the growing season they have at their disposal a considerable volume of materials from the underground reserves. But a rich supply of organic and mineral materials supports not only an intensive growth of the bines and shoots but also hastens their biological ageing. Biologically aged shoots form branches, which produce a smaller total number of cones, but these, consequently, have a bigger mass.

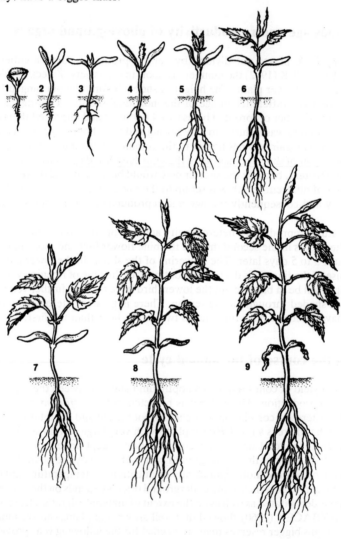

Fig. 48. Generative propagation – development of the hop seedling (1–9).

Under Czechoslovak climatic conditions a retarded sprouting of the hop plant has been shown to be advantageous. This retardation is achieved either by the mechanical (cutting, harrowing) or the chemical (diquat) destruction of the first spring shoots. Four bines for training are then selected from the resultant shoots of the retardation. This is a rejuvenating intervention which favourably affects the number of blossoms and cones. In the hop-planting region of Žatec (northern Bohemia) a minimal level of retardation is successful. The tying-in bines is retarded by one week at this level. Such a retardation encourages a large amount of top growth, increases the yield and improves the quality of cones and their number. More severe retardation was not successful because it produced smaller bines and thus lowered the coefficient *KRP*.

The rejuvenation can also be supported during the growing season, before and during blossoming and during the growth of cones, by applications of nitrogen contained for example in urea. This will prolong the vegetative period and will consequently shift the date of technical maturity. But the total number of cones will be higher and their loss during the blossoming and ripening period will be lower.

THE ECOLOGY OF THE HOP

The ecology of the wild hop and the agro-ecology of the cultivated hop deals essentially with the mutual relationships between a living organism and its environment. In the wild hop, these relationships concern the natural environment, but with the cultivated hop they involve the cultivated environment in the hop-garden. In order to obtain knowledge of ecology, both aspects require careful study of the hop plant as well as its environment. Ecological knowledge concerning the wild hop and its environment serves as the starting point for the agro-ecology of the cultivated hop.

Ecology of European wild hop

It is most important to determine the original site of the European wild hop (*Humulus Lupulus* L. ssp. *europaeus* Ryb. var. *spontanea* Ryb.) in order to study its ecology. Previous investigators assumed that area of the origin is that area where the hop occurs in greatest quantity. According to ŠREDER (1895) LEDEBURG found the greatest concentration of wild hop in south-east Russia, GMELIN said the area was in Siberia up to 62°N and PALLAS nominated the Ural and Altai mountains. These areas conform with those in the later investigations of RODNOV (1935), who mentioned that the most frequent occurrences of hop were in Siberia, the Far East, the Altai mountains, the western side of the Ural mountains, as well as in the Northern Caucasus and Kuban river regions. In addition, according to NEČIPORČUK (1955), cordatifolious hop is found in Siberia and the Far East.

VAVILOV (1936) investigated the heredity and variability of botanical species in various geographical centres and found five original genetic centres of agricultural crops. Therefrom they spread out to twenty secondary centres where the original genes were either modified by mutation and hybridization or new ones appeared. VAVILOV's important discovery forms the basis for the law of homologous series, according to which certain characteristics of plants are repeated under suitable conditions. Many ecotypes came into existence in these secondary centres and led to the so-called agro-ecotypes following cultivation. Hence cultivated varieties were bred. The genetic heterogeneity, shown by the subspecies of wild hop, together with the directions of their migration show that the secondary genetic centre of the climbing hop (*Humulus Lupulus* L.) is in Altai mountains and in southern Siberia. The wild climbing hop spread out from this centre by natural migration and by artificial migration over long distances when the semi-cultivated and cultivated hop was transferred to distant localities and there became wild. The general distribution of the climbing hop is shown in Fig. 35 (see page 63).

LINNÉ (1766) thought that the hop was transferred, together with spinach and asparagus, from eastern Russia to western Europe together with migrating populations. It is interesting that the direction of transfer from the East into Europe coincides with data concerning the usage of hop for the production of beer. KAVINA (1923) claimed that Finnish and Tchud tribes were the first to use the hop as an additive to beer. This practice passed to Slavonic and . German tribes. Before the hop was used to brew beer it served in the conservation of food and beverages. The hop is used at the present time for the preservation of honey-water in different Siberian territories such as the Altai mountains. In Central Asia hop cones are used in the preparation of bread and soups. Such a preservation of food-stock by the chemical substances contained in hop cones was of great importance for nomadic tribes and this suggests that the plant expanded from Asia to Europe.

The basic geonomic and climatic requirements of wild hops were established under the conditions of the north temperate zone where the hop genus (*Humulus* L.) and the species of climbing hop (*Humulus lupulus* L.) passed through their phylogenesis and where the European wild hop (*Humulus lupulus* L. ssp. *europaeus* Ryb. var. *spontanea* Ryb.) continues to grow in the neighbourhood of broadleaved shrubs and trees.

The climatological requirements of the wild climbing hop are those of the climate of the temperate zone including a long summer day and a moderate temperature, but also some of those of a continental climate especially as regards suitable stocks of broadleaved shrubs and trees. The accommodation of the hop plant in the climate of the temperate zone is evident from its requirements for minimum, optimum and maximum temperatures for the germination and growth of its vegetative organs.

Of the various climatological conditions typical of the temperate zone, hop is particularly *adapted to the utilization of the long summer day*. The largest increments of bine growth have been found in well-grown hop plant, under favourable temperature conditions during June, when the day is longest in the northern hemisphere.

Two thirds of the total length of bine and shoots grow during this month. Bases of inflorescences (catkins) become easily evident at the beginning of July, but their germs appear earlier. With shortening of the days in August and September, syncarpies (cones) ripen quicklier and the transfer (migration) of reserve materials from above-ground to the underground organs is done with higher speed. Therefore later produced inflorescences give smaller cones and very retarded blossoms produce no cones. This appears in the case of seedlings as well as late sprouted hop shoots.

Great effect of stock climate on the requirements of wild hop was found by MER-KENSCHLAGER (1943). According to him, it is possible *to distinguish two periods in requirements for light or for illumination*, namely:

a) shadow period which is important for growth of vegetative organs during vegetative period of the plant;

b) period of full light which is necessary to proceed into fructification phase.

Differences in the illumination affect also the temperature conditions. Under shadow, the mean temperature is lower and the day and season variations are smaller. The mean temperature is higher in full light, the day and season variations are however greater.

As to the ecological soil property, the wild hop is *hydrophylous*. Hop thrives in humid regions. If it grows wildly in an arid territory, then only on sites sufficiently supplied with water, i.e. along riversides or in inundated forests. Also the requirements of hop for *the content of humus* in the soil becomes higher, as it follows from the properties of upper layers in the stocks of shrubs, trees and perennial plants, where the mass of fallen leaves produces such conditions.

According to the ecological classification of soil as elaborated by KLEČKA and KUNZ

(1943), hop belongs to the meadow (humidiphilous, hygrophylous) type. This classification corresponds to its requirements for humus, for low acid reaction of soil and for higher content of potassium, magnesium, sulphur and manganese.

Higher requirements of wild hop for nitrogen are under natural conditions covered by mycorrhiza.

Ecotypes of wild hop were defined in Czechoslovakia by BLATTNÝ et al. (1950). Among ecological factors governing the quality of soil and climate they followed first of all hop requirements for light temperature, humus and humidity of soil. They found all wild hop plants had high requirements for humus and humidity of soil. They found most lush plants on sites with thick layer of surface humus and well supplied with water. Investigating the requirements for light and temperature they found that hop does not grow in deep shadow of inundated leafy forests or at least does not blossom there at all. If however the plant was in shadow of shrubs only at the beginning of its vegetative period, then it grew over the shrubs and its parts exposed to light were fully fructiferous.

Four ecological types were established for wild hop in Czechoslovakia:

1. inundated forests;
2. shrubs around brooks;
3. hillsides;
4. foot-hills.

1. *The type of inundated forest* is mostly bordered by the stocks of willow and alders. The hop plants are there featured by mighty height, big leaves and mostly green bine. They have requirements for humous and humid soil and for smaller intensity of illumination.

2. *The type of shrubs in brooks* is mostly featured by plants with big leaves, but the brook-stocks do not deliver such deep shadow as the inundated forests. But the requirements for illumination and its intensity are higher. Requirements of these plants for humous soil and underground humidity are remarkably high but their adaptability to variations of ground water table and to temporary lack of water is higher. They climb around willows and shrubs growing on sides of brooks, rivers, ponds and other water sources. Their bines are both red and green.

3. *The type of hillsides* occurs on drier localities together with smaller shrubs, blackthorns and eglantines. The plants are featured with smaller and finer leaves. The colour of bine, the shape of cones and their scent considerably varies. This type likes light and temperature and has high requirements for humus and humidity of soil.

4. *The type of foot-hills* occurs in higher localities (e.g. in Šumava mountains to 900 m) mostly together with low shrubs, e.g. with eglantines. These plants have small and scabrous leaves, minute cones growing on short branches from fertile shoots. Pigmentation of bines, shoots and costae of leaves is mostly red. The plants like light and have higher adaptibility to lower temperatures and lower content of humus and humidity in the soil.

As VÁGNER et al. (1974) stated by ecological investigation of wild hop done in Yugoslavia, the hop participates in microassociations with 14 to 23 per cent. As to other components, mostly trees, willows (*Salix* L.) participated with 7–29 per cent, shrub, mostly dogwood (*Cornus sanguinea* L.), golden elder (*Sambucus nigra* L.) and others with 4 up to 28 per cent. Lianas participated with 8 up to 39 per cent. The participation of plants was the lowest, namely 7 to 20 per cent. The most representative species here were stringing nettle (*Urtica dioica* L.) and bedstraw (*Gallium mollugo* L.).

It is interesting that the dominant trees, shrubs, climbers and herbage plants were similar in all of the large regions (macroregions known from Yugoslavia) mentioned. This confirms that the local soil and ground conditions and the composition of the mixed bush and tree stand has a deciding effect on the presence of wild hop. The herbage and climbing plants require the same conditions as the wild hop and therefore become its companions.

The ecological investigation of the wild hop indicates that its fertility is affected by the length of day-light and especially by the intensity of illumination. In nature, the wild hop is usually shadowed during its growing period by bushes and trees, and it is only those above-ground parts which are intensively illuminated which pass into fructification period. Shading retards those changes involved in ageing and so delays the maturing of the above-ground organs. Therefore the wild hop matures later than the cultivated varieties which originated from it.

In flooded forests where there is constantly insufficient illumination or in high lands with a short growing season the hop never flowers and is propagated only vegetatively. The basic geonomic and ecologic requirements of the hop are found under natural conditions around foliaceous trees and shrubs. The requirements for a friable soil, for adequate water and nutrients and for slightly acidic soil reaction are the most important.

Fig. 49. Vegetative propagation of wild hop (mother plant on right side).

Ecology of the cultivated European hop

The ecological requirements of the cultivated European hop have the same main features as those of the wild hop. The agroecotype of the cultivated hop most probably developed from the ecotype of a shrub growing near water or perhaps from an intermediate ecotype between this and one growing on an open hillside, i.e. from an ecotype requiring intensive illumination during the fructification phase.

Some of the external conditions in hop-gardens are different from those in the wild and scientific data are needed to determine these differences, as they apply, for example, to varieties in the genetic group "Žatecký červeňák".

The seasonal pattern of *solar radiation* in the hop-garden in central Europe is one c l i m a t i c c o n d i t i o n that is different from that in the wild. Young shoots and fixed bines receive high levels of illumination in the spring, later the plant stand becomes dense as the bines grow and ramify so that the intensity of illumination is gradually reduced. An insufficiently intense illumination prevents the formation of inflorescences. Shaded cones ripen more slowly, than those intensively illuminated. Vegetative organs grow more quickly in shade, but their tissues are thinner and more succeptible to disease. A cultivated hop of high quality requires a mean value of solar irradiation of 1800–2000 hours per year including 1300–1500 hours during the vegetative period.

Regions suitable for hop culture have a mean annual *temperature* of 8–10 °C. The vegetation temperature constant ranges from 2000 °C to 2800 °C, but totals such as there serve only for general information, because the temperature patterns and the extremes during the whole vegetative period are important. The growth of the above-ground organs requires temperatures to be above the minimum limit 8 °C, but growth stops at a temperature above 35 °C, especially when the relative humidity of the air, is low. In some cases fine tissues can be damaged in fierce sunlight. Large variations in temperature during the spring retard growth but accelerate ripening and cause earlier technical maturity.

The hop is a *hygrophylous plant*. Its main source of water is the soil with its reserve of ground water, but the above ground organs take up water from rain or dew. Not only is the total volume of precipitation important, but also its distribution during vegetative period. A reduced volume of precipitation in April or May has no substantial effect on yield because, at this time, the demand of the hop for water is low. Most of the precipitation or additional watering is needed from the end of May to the first half of August. After this time precipitation is not required, because it retards the ripening process and consequently delays the harvest of hop cones. Dense fogs are unfavourable throughout the whole vegetative period because they limit solarization and provide ideal conditions for the spread of Peronospora.

The requirement of hop plants for water depends largely on the properties and movement of air masses in the immediate environment. The properties of the air mainly concern its relative humidity, its temperature, and its content of carbon dioxide and harmful substances (such as fly ash and sulphur dioxide). Moderate air movement is conducive to a better supply of the plants with carbon dioxide and lesser likelihood of attack by Peronospora.

The requirements of the cultivated hop for *soil conditions* are fairly precise. As it is a typical meadow plant it has corresponding requirements for individual cultural factors. Thus, the properties of the soil and the subsoil horizons down to the ground water level are of great importance. The optimum ground water level ranges between 150 and 200 cm. Likewise, the stability of ground water content is very important, because if it frequently varies then the tangled roots function unsatisfactorily. Where the underground water level rises for a considerable time, roots begin to putrefy and eventually die as the physical, chemical and biological properties of the soil deteriorate.

The requirements of hop plants for a supply of *soil moisture* are considerable. By the use of lysimetric measurement from the beginning of April to the end of October, RYBÁČEK (1977) found the water consumption of hop plants to equal 482 mm of precipitation (average for 10 years). The greatest requirements were more than 100 mm per month during July and August.

The requirements of hop plants for *nutrients,* both macroelements and microelements, are high. The consumption of nutrients varies in different periods of growth. It begins to increase in the spring and culminates during the period of inflorescence and cone formation, uptake ceases after harvest when the transport of nutrients to underground organs predominates. According to ZATTLER (1956) one part of nutrients is transported to the underground organs, another part is washed away by rain and dew and the rest remains in the decayed bines.

As well as the three main elements involved in photosynthesis, carbon (C), oxygen (O) and hydrogen (H), other macroelements essential for the hop are nitrogen (N), phosphorus (P), sulphur (S), potassium (K), calcium (Ca) and magnesium (Mg) and the microelements, iron (Fe), manganese (Mn), zinc (Zn), copper (Cu), molybdenum (Mo) and boron (B). *These elements have the following basic functions:*

a) they are inseparable parts of plant cells and their constituent parts (e.g. protoplasm, nuclei);

b) they play an important role in metabolic processes;

c) they take part in the translocation processes within biochemical reactions;

d) they affect the structure of colloids and the permeability of cellular membranes;

e) they take part in the control of physiological processes.

The individual elements have their own specific functions which become most apparent when they are in excess or deficiency. These aspects will be given attention here.

The non-metallic macroelements, nitrogen, phosphorus and sulphur, constitute the main nutrients of the hop and are taken up as anions. They become constituents of essential organic complexes such as albumins and other components of living plasm.

N i t r o g e n (N) is a basic component of all albumin complexes including the complicated nucleoproteins. The high activity of albumins is implied by the chemical activity of nitrogenous complexes of amino acids. Nitrogen is present in amides and amines, in chlorophyll and in various other proteinaceous nitrogen compounds. Nitrogen promotes the growth of hop as it does that of other plants. In cases of nitrogen deficiency the hop plant becomes stunted and its leaves are thinner with narrow lobes and are of a light green colour. The cones are small and undeveloped. Where there is severe deficiency the plants are dwarfed, their leaves are minute, light green to yellow and are shed early because their growth ceases very early in the season. If there is plenty of nitrogen, the plants grow luxuriantly, the leaves are large and are deep green in colour. If the cones are present in small numbers, they become oversized and often proliferate (Fig. 50); their gross structure leads to a reduction in their valuable properties. Their axis is thicker and they contain less lupuline and have a poorer scent.

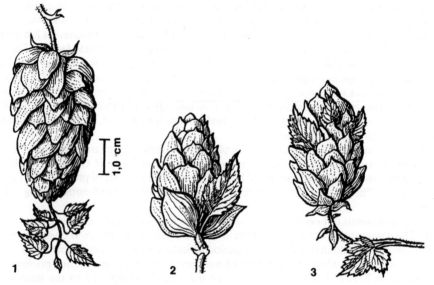

Fig. 50. Proliferation in hop cones: 1 – proliferation at the apex, 2 – proliferation at the base, 3 – proliferation in the middle.

If there is abundant nitrogen, the plant tissues become watery, flimsy and disposed to infection and mechanical damage. A graded series of nitrogen doses causes increasingly prolific growth and prolongs the individual phenological periods, thus prolonging the total vegetative period. Abundant nitrogen limits the elongation of roots and, consequently, the establishment of a strong root system in young hop plants.

Hop plants grow very vigorously, hence they need a considerable quantity of nitrogen, as do all nitrophilous plants. In natural plant stands of wild hop the nitrogen supply is supported by mycorrhizae. In the cultivated hop garden applications of nitrogenous fertilizers are needed.

P h o s p h o r u s (P) is an essential component of many organic complexes in plant cells, some of which participate in biochemical processes concerned with the transfer of energy. Phosphorus promotes the development of the generative organs and therefore at certain times its effect is the opposite of that of nitrogen. It causes an increase in the number of inflorescences and suppresses the proliferation of cones. A deficiency of phosphorus inhibits root growth and that of other underground and above-ground systems because it limits the transformation of sugars to organic acids which are the acceptors of amine groups. Thus the production of those nitrogenous substances necessary for growth becomes limited. An abundance of phosphorus, plus other factors causes premature blossoming and accelerated ripening of cones, with a consequent total reduction of vegetative period (see colour Fig. I). In phospohorus deficiency the number of inflorescences is small, the total number of cones is decreased, and the cones develop very badly, they are minute, their quality is lower and at the stage of technical maturity they do not become tightly closed.

S u l p h u r (S) , like phosphorus is a constituent of certain proteins, amino acids (cystine, cysteine, methionine), enzymes, vitamins and essences as well as many other substances. In the hop, it also affects the production of essential oils in the cones. The morphological symptoms of sulphur deficiency are similar to those of nitrogen. They are mainly seen as retarded plant growth and lower synthesis of chlorophyll so that leaves are pale green and they soon decay. Hop plants are susceptible to damage by inorganic sulphur compounds, particularly sulphur dioxide and hydrogen sulphide, which occur in the environment of large industrial agglomerations.

The metallic macroelements, potassium, calcium and magnesium have a special place in the hop plant. The plant takes them up as cations the function of which in metabolism is different from that of anions. They participate in different control processes in cells, tissues and even organs. *They are found in plant cells in three different forms as follows:*
   a) ions (in vacuoles, in the solution within the xylem and in cytoplasm);
   b) ions in adsorption bonds (in the structural proteins of the cytoplasm);
   c) fixed structural units of organic and inorganic compounds.

P o t a s s i u m (K) serves in various types of metabolism. It participates in the metabolism of glycosides, amino acids and proteins. It is active also in the metabolism of phospohorus and in the activation of enzymes. It has a marked effect on the activity of plasma colloids. It increases the strength of plant tissues and thus increases their resistance against infection and other forms of damage and thus improves the general health of hop plants. It has a positive effect on the ripening of hop cones, the production of lupuline and the content of resins and essential oils. Potassium deficiency is seen mainly in old leaves because the young leaves use potassium drawn from older tissues. Affected leaves turn pale, starting from the edge and later develop brown spots bordered by the veins. These spots gradually increase in size, the leaves become brass yellow and finally turn ash grey before they are shed. Potassium deficiency upsets the apical dominance of the bines, which produce branches and these are then longer than the leader. An abundance of potassium has a negative effect on the utilization of other ions (particularly magnesium), causing a deterioration in the quality of the cones which have a lower content of lupuline and resins.

C a l c i u m (C a) is found in plant cells in solution either in water or in acids but it forms insoluble as well as soluble compounds. It forms salts with mineral (carbonic, sulphuric, and phosporic) and organic (e.g. pectic and oxalacetic) acids. The water soluble phases are found in the protoplasm, bound to the acidic groups of proteins. They are mobile. The acid-soluble phases occur in the vacuoles and are immobile. Calcium plays an important role in the production of cellular membrane and is an important part of many cellular contituents. The processes of cell division and cell elongation require its presence. Because calcium is not very migratory and tends not to be re-utilized, its uninterrupted supply is required throughout the whole vegetative period. Where calcium is deficient, the cellular membranes become muciform and tissues tend to premature lignification. This deficiency first appears in the youngest organs, at the growing point, and in young leaves. The plant apex becomes yellow and dies. The apical leaves are minute, convex with a pale edge. Later brown spots appear on the surface of the leaves, which then decay. An abundance of calcium reduces the uptake of the cations magnesium, potassium and iron and this can result in chlorosis. Abundance of calcium also causes the structure of cones to become coarse and they turn yellow prematurely.

M a g n e s i u m (M g) *is present in plant cells in three forms:*
a) bound by protoplasm, in the chlorophyll molecule or vacuole;
b) in organic salts;
c) free.

Only 10 per cent of the total amount is bound in the chlorophyll where it has a specific function in photosynthesis. In addition, magnesium strongly affects the production, transformation and transport of glycosides as well as the synthesis of fats and proteins. Magnesium also has influence on the formation of reproductive organs and – in the hop plant – on the quantity and quality of hop cones.

Magnesium raises the amount of lupuline in the cones and improves their green coloration. It can be re-utilized, and so it moves from leaves to reproductive organs as required. Magnesium deficiency first appears on old leaves as chlorosis. The leaves first become pale, then become yellow with green strips along both sides of the veins. The chlorotic areas later turn grey, becoming red-brown and the leaves fall prematurely. An abundance of magnesium occurs only rarely.

Microelements occur in hop plants in only small quantities, but some of them are indispensable and highly efficient. These include iron (Fe), manganese (Mn), zinc (Zn), copper (Cu), molybdenum (Mn) and boron (B). The importance of the microelements increases with larger doses of mineral fertilizers, especially on permanent hop plantations in traditional hop-growing regions. *The main functions of the microelements are as follows:*
a) they act directly on the physiological and biochemical processes in plant tissues;
b) particular microelements work very specifically on particular processes;
c) they act very efficiently even in very small concentrations;
d) when they are deficient the normal life cycle of the plant is disturbed.

An abundance of microelements has the same pernicious effect, as their deficiency, because high concentrations are toxic. Microelements occur in plants in different forms. They mostly form complex compounds with organic substances or organomineral complexes with mineral and organic substances.

I r o n (F e) is mainly concentrated in the leaves and up to 80 per cent of the total iron is present in chloroplasts. The reason for this is that the iron is directly involved in the chlorophyll system, but it is not retained there. A deficiency of iron appears first on young hop leaves as a typical chlorosis.

M a n g a n e s e (M n) is mainly concentrated in hop leaves and seems unlikely to be available for re-utilization. The amount present in the plants increases with increasing plant age (PICHL, RYBÁČEK, 1972). Manganese deficiency slows the growth of bines and branches, shortens the blossoming period and prolongs the ripening of cones, thus delaying technical maturity.

Z i n c (Z n) is mostly concentrated in apices and young organs, such as young leaves. Its amount decreases as the plant ages, but zinc is transported and re-utilized in young organs. Leaves become brittle when zinc is deficient as is described in the section on harmful agents (see page 131).

C o p p e r (C u) according to the investigations of PICHL and RYBÁČEK (1972), is present in the above-ground parts of the plant. It is distributed like zinc and can also be re-utilized. The effects of a deficiency in copper cannot normally be described because in Czechoslovakia copper, and various its ions are applied as protectants against diseases of the hop plant.

M o l y b d e n u m (M o) – symptoms of deficiency have been described in the literature, but they have not yet appeared under Middle-European conditions. They become apparent on young leaves which are chlorotic with light brown spots around the veins. Later they turn white, their edges roll up and generally become necrotic and fragile. Development of these symptoms grows with the age of leaves, and when there is a severe deficiency of molybdenum they involve the whole hop plant.

Boron (B) cannot be re-utilized and therefore its presence in leaves increases with age. When boron is deficient, growth ceases, the shoots become thickened and rigid, and the internodes are shortened. The shoots, while still young begin to decay from the apex. Leaves are small and often deformed.

Critical values for the content of certain microelements in hop leaves are shown in Table 26.

TABLE 26 **Critical values for microelements in hop leaves**

Microelement	Content in leaves (mg per kg)				
	insufficient	low	medium	high	toxical
Manganese			0.37 – 1.14		> 2.320
Molybdenum	< 0.04		0.04 – 1.03	> 1.03	
Zinc (in crop)	< 12	12 – 61	62 – 120	> 120	

With regard to other soil components, hop has a high requirement for humus content, which aids the better sorption of mineral nutrients. Its organo-mineral complexes have 6–7 times higher sorption ability than that of the argillaceous minerals themselves.

A weak acid reaction (pH 5.6–6.5) is the most suitable for the hop although it can tolerate neutral soils (pH 6.6–7.5).

Ecological conditions in Middle European (Czechoslovak) hop planting regions

In the Middle Ages, hops were planted in areas near convents, monasteries, towns and villages over almost the whole of the territory of the present Czechoslovak republic. Later, the areas became concentrated in certain hop-growing regions. This was stimulated by the requirements of breweries for a certified provenance of their hops. As early as the 1500s particular West Bohemian towns such as Rakovník, Žatec, Louny, Beroun and Klatovy were given written certificates and a seal. These ancient rights of particular towns were improved and unified by imperial decree issued by Maria Theresia in 1769. The permanent hop-growing regions became clearly established definite form during the second half of the 19th century as the first hop-token office was established in the town of Žatec in 1884. A Bill in 1921 defined five hop producing regions, namely the regions of Žatec, Úštěk, Roudnice, Dub and Tršice and introduced compulsory tokening of hops produced in these regions.

Legal measures in 1962 confirmed certain facts, such as the obsolescence of the region of Dub since 1945, the linking of the region of Roudnice with that of Úštěk in 1957 and the establishment of two new regions namely the region of Piešťany and Topoľčany in the west of Slovakia, consisting of 39 localities in districts Topoľčany, Trenčín, Galanta and Trnava. The precise determination of the hop-producing regions on a compact territory, involving villages and communities is the most rigid territorial system of plant production in Czechoslovakia if not the whole world.

The largest region is that of Žatec with 355 planting villages in the districts of Louny, Chomutov, Kladno, Rakovník, Rokycany and the northern surroundings of Plzeň, next is the region of Úštěk with 220 villages in the districts of Litoměřice, Česká Lípa, Mělník and Kladno, then the region of Tršice with 65 villages in the districts of Olomouc, Přerov and Prostějov, the region of Piešťany and Topoľčany has 39 villages in the districts of Topoľčany, Trenčín, Galanta and Trnava.

Fig. 51. The areas planted with hops in particular districts of Czechoslovakia.

Climatic conditions in the Czechoslovak hop-growing regions

Czechoslovak hop growing regions are situated where the climate of the maritime temperate zone meets that of the continental zone. Moreover, the Žatec region is in the rain shadow of Krušné hory (Ore Mountains) and Český les (Czech Forest) and this gives the climate of this region certain special features.

The whole complex of climatic conditions is made up of individual factors, including temperature, illumination, moisture and air movement. Abnormal variations in any of these factors not only affect the weather pattern but also change the local climate and microclimate in the hop garden and thus affect the individual hop plants. Therefore, knowledge concerning abnormal variations of individual climatic factors is most important in hop-growing.

The hop, under Czech conditions, is *resistant to winter freezing,* but it is very *sensitive to spring frosts,* which not only totally stop the growth of the bine but will also damage young leaves if the temperature is lower than −5 °C.

For generative propagation, i.e. for the growth of inflorescence and cones, the hop plant requires *intensive illumination.* Under insufficient illumination, the development of the inflorescence is loose and retarded and consequently the flowers are small in number or not appear at all. The cones which develop in the shade, grow slowly and produce only a small amount of lupuline. In contrast, the growth of other organs in shadow is accelerated but the tissues of poorly illuminated above-ground organs are soft and disposed to become diseased. Czechoslovak hop-growing regions, particularly in the region of Žatec, have little cloud during the blossoming and ripening period and a high intensity of solar radiation, which favours the growth of good quality cones.

The requirements of the hop plant for *moisture*, as well as other factors, has a great effect on growth and development and in consequence moisture controls the quality of hop cones. Increased atmospheric humidity and increased precipitation during the growing period is favourable so long as it does not produce conditions suitable for attack by *Peronospora* or does not inhibit the use of agricultural machinery. Most Czechoslovak hop regions have a small total amount of precipitation. In the town of Žatec, which is situated in the centre of the most famous region, the mean annual amount of precipitation is 449 mm. This amount is not sufficient for the optimal development of hop plants nor to achieve their maximum growth. Therefore, in the selection of land suitable for hop gardens it is necessary to choose localities where the plants can, if necessary, be provided with water from the subsoil resources from river valleys and creeks.

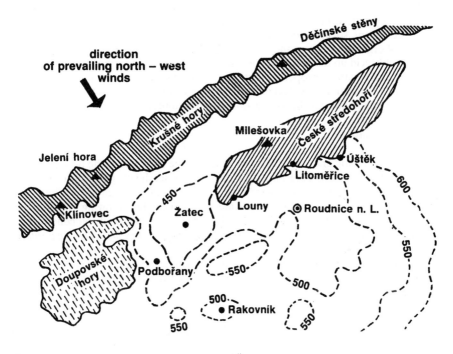

Fig. 52. Precipitation in the hop-growing region of Žatec.

An additional source of moisture for the hop is *overnight dew*. Hop plants take moisture not only through their roots but also through above-ground organs, particularly the leaves. ZATTLER (1982) found that one hop hill can take 0.6–1.7 l dew during one night. According to these data it would seem that in Czech hop gardens dews can provide up to one fifth of the total requirement of water.

In critical periods, when natural precipitation and dew are not sufficient, it is useful to supply water by *irrigation*. Here, it is necessary to consider the seasonal requirements of hop for moisture. The requirements are maximal during the intensive growth period of the above-ground organs. However, the use of excessive amounts of water or watering at the wrong time could damage the fine structure of the cones.

Soils in Czechoslovak hop-growing regions

In Czechoslovak hop-growing regions, there are different types of soils (the steppe black soil or chernozem, the humus-limestone soil or rendzina, brown soils etc.), each of them in different varieties (loamy, sandy, clay). These soils have developed on different petrographical and geological bases. In *the hop-growing region of Žatec,* the hop gardens are mainly on soils which originated from Permian geological formations. These soils, called permian red-earths (krasnozem) contain significant amounts of iron compounds (6–7 per cent of iron trioxide) and manganese as well as compounds of other metals.

The eastern part of *the hop-growing region of Úštěk* is situated on a tertiary cretaceous system and its central part, the so-called Bogland of Polepy, is on quaternarian sediments, whereas numerous igneous basalts are found in its western part.

The hop-growing region of Tršice has quaternary and partially tertiary soils. The permian red-earths of the Žatec region are regarded as producing the best hops. These soils are mostly loamy and clayey and, after deep ploughing, have a high air-, and water-holding capacity as well as a significant sorption of geoponic nutrients. However, they require systematic applications of organic fertilizers and sometimes need to be limed. The best soil reaction for the hop is weakly acid up to neutral. So, whether a soil is suitable for hop growing is not only dermined by their natural properties but àlso, to a high degree, by the quality of their cultivation and preparation, the level of fertilizer applications and such other interventions as provide favourable conditions for the growth and development of hop.

Situation of hop gardens

As well as the climatic and soil conditions, it is also the situation of the hop garden which affects hop-growing. This is determined by height over sea level, position in the relief of country, the slope of the terrain towards the cardinal points. Czechoslovak hop-growing regions are geomorphologically very complicated. Within any hop-growing region, there are various valleys and other local situations which affect the climatic and soil conditions.

According to the position of the hop garden in the relief of country the following positions are fixed:
1. in the inundation zone within the flood territory of rivers and creeks;
2. downstream in valleys of rivers and creeks, outside the inundation zone;
3. on the hillside above the valleys of rivers and creeks;
4. dale gardens on the flat lands between hillsides;
5. on open table lands.

Flat lands in valleys where there is free air movemont but well-protected against northern

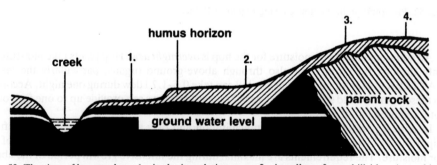

Fig. 53. The sites of hop-gardens: 1 – in the inundation zone, 2 – in valleys, 3 – on hillsides, 4 – on higher land.

and western winds are the most suitable for hop-growing. Open field positions may be suitable only if the hop garden is protected by surrounding trees.

The best sites are in the valley of Zlatý potok (Golden creek), in the region of Žatec, at the locality called Podlesí (Woodland) and in the region of Úštěk, at the locality Polepská blata.

Sites in narrow or closed valleys are not suitable for hop production because the air circulation is poor and is conducive to fungal diseases. Totally exposed, unprotected sites in fields are also unsuitable because the plants are likely to suffer wind damage. Gardens in the inundation zone are susceptible to occasional flooding and in spring the hop-garden structures in such sites can be damaged by ice floes.

Experience over the past hundred years or more showed that choice-quality hops can be developed only under favourable conditions, such as those in Czechoslovak hop-growing regions. These requirements must be respected when sites for new hop-gardens are selected, when the microclimate in a hop garden is in any way modified or when hop-land reclamation is being used to provide hop planting sites. The effects of soil, climatic and ecological conditions on the quality of cones was discussed by ZELENKA (1965) who gave *the following evaluation of natural conditions in Czechoslovak hop-growing regions:*

1. low-land regions, i.e. the warm parts of hop-growing regions, where hops are produced with high yield and a high content of bitter substances –
   a) on sand-loamy and loamy soils where there is high yield but a lower quality of cones;
   b) on clay-loamy soils with middle to high yields and high cone quality;
2. high land regions, i.e. cold parts situated –
   a) on sand-loamy and loamy soils with low to middle yields and high cone quality;
   b) on clay-loamy soils, red-earth with low to middle yields and exellent cone quality.

IDEAL AND ACTUAL TYPES OF HOP PLANTS AND THEIR STANDS

A stand of the cultivated hop is formed by individual plants regularly distributed over the area of the hop-garden. The hop garden itself is an independent agricultural unit which exists as part of the national land reserves for a fixed period, usually 20–30 years. During the whole period of its existence the hop garden is affected by the environment, i.e. by geoponic and climatic factors, by the biological properties of individual hop plants and it is essentially controlled by those agrotechnical factors which affect both the environment and the plants.

The basic characteristics of all three groups apply at the time when the hop garden is created. The geoponic conditions depend on the choice of ground and on its previous reclamation; the biological properties of the plants depend on the choice of variety and on the quality of the planting material. Among the agrotechnical measures applied during the creation of a hop garden, the depth and method of planting has a permanent effect. The height of hop garden structures is very important as it affects the habit of the above-ground parts of the hop plant. If the bine growns beyond the top wires its apical dominance is disturbed and this affects branching. Differences in the habit of above-ground plant parts affect the local microclimate in hop gardens especially with regard to light, temperature and humidity. The ideal type (idiotype) of plants and stands of hop depends on the qualitative requirements for the idiotype of the commercial product, i.e. the technically mature cones.

Idiotype of high quality cones

This idiotype was developed in Czechoslovakia at about the beginning of this century. The basis for its formation was mentioned by CHODOUNSKÝ (1902) and OSVALD (1928) in their studies. Their data were used to develop a m e a s u r e   o f   c o m m e r c i a l   v a l u e which is now included in the Czechoslovak state standard ČSN 46 2520. This system involves three different evaluations in which it is possible to reach following maximum totals of points:

sensoric (organoleptic). .50 points,
morphological (mechanical) . . . . .20 points and
biochemical (chemical) . . . . .30 points.

The ideal type of high quality cone would obtain the highest number of points in each evaluation to give a total which is maximally 100.

Another system of value is used in the breeding of hop whereby, in each group, the maximum number of points is 20, i.e. 60 points altogether, and 40 points (maximum) serve for evaluation of the yield. There are *two organoleptic indicators:* the structure and scent of cones. A desirable structure is determined visually by the appearance of the cones. This structure applies to cones with symmetrically positioned and alternately overlapped cuticules and bractlets; in a perfect cone, the axes are placed exactly half way between the axes of neighbouring cuticules and bractlets.

The highest number of 10 points is obtained by cones 25–30 cm long. Shorter and longer cones lose 1 point for each 5 mm difference from the optimum. Shorter cones are usually not fully developed and longer ones are overgrown sometimes with proliferating small leaves. If the ratio of length to width of cones is 3:2, the cones are of elongated oval in form and their dimensions are ideal. The maximum number (10 points) is given to cones with the required fine hop scent typical of the variety 'Žatecký'. Less definite or a sharper and more penetrating scent is given 8 points. Cones with a sharp spicy scent or with a vague non-typical hop scent obtain 6 points. Cone with undesirable scents (e.g. garlic, fruit etc.) which hide the proper hop scent and cones with a low or non-existent component of hop scent receive only 1 to 4 points.

With regard to *the mechanical analysis*, it is the percentage of strigs (part of axis of cone) in the total mass of cones and the regularity of strigs which is used as a measure of the quality of hop cones.

The main indicators of *chemical quality* are, at present, the content of α-fraction (humulone fraction), β-fraction (lupulone fraction) and polyphenolic substances (tannins). The highest rating is given to hop cones with a content of α-fraction over 5 per cent, β-fraction over 7 per cent and the ratio of both can vary from 1:1.3 to 1:1.5 (with the average 1:1.4).

The variety 'Žatecký' most nearly approaches the idiotype of high quality cones. According to VENT et al. (1963) the mean length of a cone was 27.84 mm and the mean width was 18.51 mm, giving an almost perfect required ratio of length to width (3:2). Likewise, the mechanical analysis approached the required idiotype of a top quality cone. The percentage of spindle was an average, 8.6 per cent, a typical density of the spindle ranged from 6.30 to 6.70 and the mass of cones from 1.09 to 1.27. The average mass of 100 dried cones was 15.0–16.5 g, what corresponding to a fresh cone mass of 60 to 70 g.

Idiotype of the hop plant

The idiotype of the hop plant links the idiotype of the cone with the productivity of the hop plant. According to the methods for the evaluation of cultivated plants used in the Research Institute of Hop Culture in Žatec 40 points of a possible 100 relate to the yield and the other 60 to the quality of cones.

The fixed idiotype of a top quality cone together with other balancing factors predetermines

114

to a large extent the idiotype of the above-ground part of the hop plant and largely governs its main properties.

Because the idiotype of the quality cone governs its weight (mass), the yield of hop plants depends on the number of cones it produces. The various characterictics of cones as well as the yield of plant are highly affected by the shape of the above-ground parts of hop plant.

Fig. 54. Growth habits of the above-ground parts of hop plants: 1 – umbrella (hat-type), 2 – cylindrical. 3 – conical.

There are *three basic types of growth habit,* namely conical (steeple), cylindrical and reverse cylindrical (umbrella) with intermediate phases. In terms of balanced cone production, the best habit is conical, but it produces a very low yield of cones because of the short length of side branches. The umbrella type by contrast, has top branches sometimes over-growing and shadowing the lower ones so that this growth form not only produces a lower yield but its cones are of lower quality. In terms of overall yield and quality of cones, the best habit is cylindrical which has approximatively same length of fertile branches in the central and top part of the whole plant.

The idiotype of the above-ground parts of the hop plant not only affects the size and properties of the underground parts but also controls the following items:

1. number of fixed bines on one plant;
2. number of bines on one hop-pole;
3. number of fertile branches on bine;
4. mean length of fertile branches;
5. amount of further branching of fertile branches.

The growth habit of the above-ground parts of the plant is naturally affected by the number of bines fixed on one hop-pole. If three-bine fixing used, the above-ground part is likely to have a greater weight (mass), but relatively lower ratio of cones, i.e. a lower reproduction coefficient (RYBÁČEK, 1975). One-bine fixing in contrast often produces cones that are less equilibrated and are often overgrown. Therefore, well-grown hop stands produce the best results with two-bine fixing giving mean values of reproduction coefficient and a mean equilibrity of cones.

The requirement of a maximum number of fertile branches is no less important. The fertility of all branches is a genetically fixed property of the idiotype of the hop-plant. If light intensity in the hop-garden is limited, branch fertility will not be fully realized and the cones will grow only on the top most branches.

The number of fertile lateral branches is usually determined by the number of nodes between the first pair of fertile lateral branches and top wire of the structure. The apex has a large number of short nodes and can show intensive branching but the branches do not develop.

The number of fertile branches can also be estimated according to the ratio of the fertile to the infertile part of the bine but this estimation is less precise. The infertile bine is that part from the soil surface up to the node from which the first fertile branches develop, above this is the fertile bine up to the top of the structure or up to the apex of the bine if it does not grow as far as the top wire.

RYBÁČEK (1967) found, on different variants in production stands that there is a negative correlation between the number of bines on one hop-pole and the height of the structure, and consequently of the number of fertile branches. The number of fertile branches on one bine decreases with increasing number of bines per pole.

It is time-consuming to determine the average length of fertile branches, but it is a simple matter to find their weight (mass). Given the average weight (mass), the average length of one branche can then be calculated. The weight (mass) of fertile branches is also a basic datum to establish the amount of bifurcation of branches. This figure can be determined from the relationship between the weight (mass) of branches and number of cones.

RYBÁČEK (1967) found that there was also a negative correlation between length of branches and the intensity of their bifurcation. Long branches have fewer bifurcations, whereas short branches bifurcate very intensively. The largest number of cones can be found on medium sized branches ranging from 50 to 70 cm, growing on a hop plant of cylindrical habit.

Short branches, the number of which is higher if the habit is conical, bifurcate intensively, but this intensity cannot compensate for the insufficient length of the branches and the consequent small number of cones. Numerous long branches with the reverse conical (umbrella) habit bifurcate very rarely or not at all. The potential number of cones is proportional to the intensity of bifurcation of the branches.

The amount of bifurcation affects the main yield element of the hop, i.e. the number of cones and the mean mass (weight) of a single cone. RYBÁČEK (1967) again found a negative correlation between these two elements. A nearer approach to the determined idiotype of the cultivated cone does not make it possible to compensate for a lower number of cones by their higher mass (weight) and therefore all attempts to achieve the ideal type of cone must aim towards obtaining the highest number of cones.

The ideal hop plant, especially as regards its above-ground part, requires carefully thought out cultivation procedures. A study of the dynamics of the yield elements of the ideal hop plant will help to decide on most efficient agrotechnical measures. These parameters also need to be known if a forecast is to be made of the next harvest of cones.

Idiotype of the stand

The idiotype of the stand integrates the idiotype of the cone and of the plant and the production of cones per unit area of the stand (e.g. 1 ha). Hops are planted in a precise pattern and therefore it is possible to correlate, reliably, the indicators of the actual, as well as of the ideal type with the area of the stand. In this context, interest is focused first on the production of cones from the area of the stand. This production depends on the amount of above-ground bio-mass and the ratio of cones to this biomass, which was designated by RYBÁČEK (1975) as *reproduction (propagation) coefficient of the stand (KRP)*.

The amount of biomass depends mainly on the number of fixed bines per unit area of the stand, on the chronological and biological age of under-ground parts of the plants and on the different environmental and ecological factors of the garden. In an established stand, the age and number of fixed bines is stabilized, and it is the external factors that are variable; some stands are not equilibrated and here the plants show a heterogeneity of age and ontogenesis.

The number of fixed bines per 1 ha of the hop-garden depends on the planting pattern and on the number of fixed bines generated from one plant, as shown in Table 27.

TABLE 27 **Effects of planting pattern and numbers of fixed bines on the density of stand on 1 ha of hop garden**

Planting pattern (canopy) (cm)	Number of fixed bines	Area (m²) for		Area (1 ha) covered by	
		1 hill	1 bine	hills	bines
120 × 120	2	1.44	0.720	6944	13 888
140 × 140	2	1.96	0.980	5102	10 204
150 × 150	2	2.25	1.125	4444	8 888
160 × 160	2	2.56	1.280	3906	7 812
260 × 110	4	2.86	0.715	3496	13 984
280 × 100	4	2.80	0.700	3571	14 284
300 × 100	4	3.00	0.750	3333	13 333

CALCULATION OF YIELD

The yield of the hop stand is given by multiplying the mean cone mass by the number of cones per unit area.

Providing yield elements

Female hop plants have a powerful ability to produce inflorescences and syncarpy, i.e. hop cones. The potential number of flower-buds depends on the number of terminal buds theorically able to be transformed into flower-buds. The average cone mass (weight) depends on the vigour of their growth as cones mature. To determine quality of hop cones the average mass of one hundred cones is used, therefore no compensation for a smaller number by higher mass is required. Therefore, all agricultural and technical measures must be concentrated on exploiting to the highest possible degree the potential for producing the greater number of cones. This may be inhibited by internal obstacles in the transformation of vegetative buds into flower buds and also such external conditions as the intensity of light. Moreover, the further development of flower buds, and the inflorescences generated from them, depends on the internal conditions of the plant as well as on external factors.

The total number of cones in the stand depends on the following gradually developing factors: number of individual plants per unit area of the stand, number of tied-in bines and number of fertile shoots per bine. The theorical number of individual plants is determined from the planting pattern, the number of fixed bines is governed by the number of hop-poles in the training structure and the number of fertile shoots, depends on the height of the structure. These basic conditions are complemented by the size of above-ground part of the hop plant and its growth habit. These two latter properties are particularly affected by the ontogenetic period of the plant and by the ecological conditions in the particular annual cycle. An efficient procedure for regulating the yield is thus dependent on the quality of knowledge concerning ontogenesis and the physiological and ecological processes operating during the current annual cycle of the plants in the given stand. The size and habit of the above-ground parts of hop plant are, at the same time, also affected by the ontogenetic period and the course of the annual cycle.

Given the same size and habit of the above-ground system, the potential number of cones depends on the average length, and the intensity of bifurcation, of the fertile shoots. These two indicators are negatively correlated. Thus, with increasing length of the shoot the intensity of bifurcation decreases. The greatest number of potentially fertile buds is obtained with an intensive bifurcation and medium sized shoots. The density of set cones per 10 cm length, or per 10 g fresh weight (mass) of shoot serves as the indicator of intensity of bifurcation. Among the internal factors, biological age has most influence on the density of set cones. Among the external ecological factors the intensity of lighting is particularly effective. With increasing biological age of shoots their tendency to bifurcation (branching) increases, as does its consequences involving the setting of flowers and cones. The intensity of branching decreases as light intensity falls. Furthermore, an additional reduction of light intensity resulting from shadowed shoots also reduces the numbers of flower buds, flowers and cones.

Compensation for reductions in yield-controlling elements

When those factors which control yield begin to offer the above-ground parts of the hop plants during the growing season, they tend to become reduced to different extents; thus, the organs with smaller mass and higher requirements for living conditions are usually more reduced. A reduction in a primary element of yield can be compensated for with an element of lesser importance.

From the biological point of view there is no reason for the number of adult plants in the stand to be reduced during the annual cycle. Nevertheless, such a reduction can occur due for example to poor management of the garden, particularly in cutting, fixing and cultivation. A loss of individual plants results in gaps in the stand. If the gaps are not filled, this reduction in plant numbers can increase and reach 20 %. Compensation for gaps in the stand can be made by the tying-in of more bines from neighbouring plants, providing that the individual gaps are small (PAUL, 1978). In a stand with long spaces in the row such a compensation is not feasible. Therefore, a gapped stand can only be improved by a systematic replanting with rooted cuttings.

The number of bines may be reduced as early as their tying-in and, later, during the growing season, because of loss from the limited number of fixed bines. A partial reduction of yield must occur if any change in the bines causes them to be weakened. Such changes could be, for example, in the inclination of the apex of bine, its destruction, or damage leading to a change in apical dominance. An undamaged bine can be assumed to have been, in some way, weakened if its yield is more than 20 per cent lower than that of a neighbouring bine on the same pole. These yield reductions associated with the training of bines, can be substantially reduced by a careful tying-in and regular inspections to correct any faults in the tying of lines and any associated problems of cone development.

It has been found that under favourable conditions the compensation of yield in one-bine fixing can reach 80 per cent of that given by good plants with two-bine fixing. The yield of cones can moderately increase in three-bine plants but their quality is always affected.

A complete reduction of fertile shoots results from such damage to the lateral buds as to cause the shoot to fail to develop or when bines are broken and pulled out. A loss of yield due to such lost development of shoots can be in part made up by an increased yield from normal shoots situated higher on the bine. Where bines broken and pulled out this form of compensation is not available.

The greatest reductions occur on the generative organs, where the ultimate result is a premature shrivelling of the cones. This is described in the chapter "Protection of the hop"

The regulation of the yield-giving process applied to hop stands

An appropriate regulation must fit the required economic aims. With regard to the ontogenesis of the hop plant, the first requirement is the earliest possible development of full fertility, and the second is to maintain this fertility for as long as possible. These two requirements involve control both of overall ontogenesis as well as each annual cycle.

Investigations by RYBÁČEK (1957) showed that it is possible to develop full fertility in the hop before the ontogenetic period of maturity. From theoretical point of view, however, full fertility cannot be achieved while there is still strong apical dominance, because this inhibits the growth of more than one bine. This applies to plants produced from seed, during the seedling period, and to rooted cuttings, obtained from seedlings and suckers, during their rooting period. This implies that an early separation of the seedling material from the maternal plant is a basic condition for an early inception of full fertility. The most suitable material to achieve this condition is given by rooted cuttings from seeds as these can be planted one year earlier. The best time for them to be planted is the autumn.

The early development of full fertility also requires formation of maximum food reserves in rooted cuttings. This is aided by increased photosynthesis in the nursery and by an improved migration of glycosides into the underground organs. Photosynthesis particularly promotes an extension of the total foliar surface and prolongs the period of its activity. The accumulation of nutrients, particularly of starches, in rooted cuttings is helped by the limited, or hindered, production of cones (in those plants derived from cuttings) and the total decay of the above-ground system under such conditions that all of the mobile substances can migrate to the underground parts. The aim in planting rooted cuttings in nurseries is mainly to promote their growth and thus increase their biomass while at the same time hindering the formation of cones.

During the ontogenetic period of young hop plants the main aim is to achieve the greatest possible increment of above-ground and underground biomass; yield of cones is of only secondary importance. Therefore the aim is to accelerate vegetative growth in order to nullify the delay caused when rooted cuttings are used for replanting. Even the cutting treatment of new wood is abandoned so as to accelerate vegetative growth. Up to six bines from one plant may be fixed to the hop poles, without regard to the size of bines, simply to enlarge the volume of biomass. Harvesting of cones is done manually in order to avoid damage to the plants and thus to ensure the transfer of all substances from the above-ground organs to those underground in the process of natural decay.

During the ontogenic period of adolescence the yield of cones gradually becomes the main aim, though this has to be associated with the highest possible support from the underground system. Therefore vegetative growth is only very moderately reduced in consequence of the autumnal cut. Subsequently, the four most developed bines are selected and tied in two hop-poles. In order to encourage a vigorous underground system it is again necessary to avoid the harvest decapitation and to do the harvesting manually.

These regulative systems are valid for plants grown from cuttings, from seedlings and other rooted cuttings. In plants grown directly from seed the regulations are introduced one period earlier, i.e. the work with them starts as early as the ontogenic seedlings period.

In order to obtain the maximum period of full fertility, it is necessary to promote its early development and to delay the onset of senility. Likewise, as senility approaches it is necessary to maintain productivity of the maximum level. One favourable factor is the natural rejuvenating process which occurs during the chronological and biological ageing of the underground systems of the hop plant. This rejuvenation can be obtained, for example by such structural changes as favouring a higher percentage of younger organs. Thus, the structure of the root system can be changed by promoting the regeneration or restitution of the old roots, and this can be encouraged by deep inter-row hoeing. This procedure cannot be applied to the old wood, because its youngest parts, i.e. new wood and new suckers are systematically removed and so the age of the old wood is maintained. The result is the formation of sectors in the old wood, and therefore an appropriate solution has to be found. The proposal by NEČIPORČUK (1955) to limit the main roots, is not appropriate because it also limits the productivity of the hop plant.

One acceptable means involves good nitrogen nutrition and watering promotes rejuvenation of under-ground parts of the hop plant. A similar effect can be obtained by encouraging material migration from the above-ground part of the plant into its underground parts. If the delaying of the ageing processes is to be successful, it is necessary to maintain good sanitary conditions for the plants. Furthermore, it will be necessary to elaborate other measures for the regulation of certain physiological processes. These, as well as basic ecological relations, are mentioned in the following chapter "Technology of production".

CHAPTER III.
TECHNOLOGY OF PRODUCTION

The technology of production involves the sum or complexity of biological, technical (including chemical) and economic elements. These basic components of the technology of production are not homogeneous and often act in opposition.

The economic approach is to obtain the maximum yield of cones, to improve their quality and to reduce the costs of production. The reduction of production costs is achieved mainly through the replacement of manual work by mechanization and the use of chemicals. Chemical treatments and particularly mechanical operations are less able than manual workers to give full respect to the biological properties and requirements of the hop, but when used properly they should accommodate the basic biological properties of the crop. In certain cases the new technical developments require a permanent genetical change in the properties of the hop plant. This is achieved by the breeding of new varieties.

Therefore, the technology of hop production involves the interrelated technical and economic requirements of the crop and the chemical treatments and mechanization concerned in production.

The technology of production is usually divided into the following sections:
1. the planting of the hop garden;
2. the cultivation of the growing hop stand;
3. the technology of harvesting and the post-harvest processing of cones.

STAGES IN PRODUCTION TECHNOLOGY

Production technology has the following stages:
1. the basic layout of the stand, as determined by the planting of the garden;
2. the cultivation and treatment of soil in the working garden;
3. the cutting (forming) of plants and the attention given to the stand.

Basic layout of the hop stand

The basic layout of the hop stand is defined when the garden is planted. It comprises the shape and size of spacing plan and the number of fixed bines from one plant. Further changes, at a later date, are very difficult and may be impossible without basic changes to the structures.

The early arrangement of hop stands was based on experience with small-scale production which indicated that the best training method was to use two bines from each plant. This led to the planting of a large number of plants per ha, with narrow spacing (see Table 27, page 117). This arrangement of the hop stand was suited to manual cultivation and met the requirements of the hop plant for both soil and air space.

The best arrangement to suit the ideal cylindrical habit of the above-ground parts of the hop plant would involve a circular planting pattern. But this pattern cannot be accommodated in a dense stand, and therefore the nearest to a circular pattern is equilateral triangle (see Table 27). This spacing allows each plant in the stand the same distance from its six nearest neighbouring plants. The main advantage of triangular spacing is better illumination of the bines, because the individual plants in each row are situated in the centre of the space between two neighbouring rows. The disadvantage of this spacing is seen when the tractor is used during the cultivation of inter-row spaces. When triangular spacing is used the distance between rows is less than the distance between neighbouring plants within the row. This spacing was not suitable in the days when horses were used to pull the cultivating implements.

The use of tractors was made possible due to a new design of the spacing surface with larger dimensions. Tringular spacing was replaced by quadratic spacing with two bines of each hop plant fixed to one pole. Thus, each hop plant in the stand had 4 adjacent plants as its immediate neighbour. This arrangement approaches the original ideal circular pattern namely as compared with that of regular hexagon in triangular planting. In this quadratic spacing, the distance between rows is the same as the distance between the plants in the rows. As the distance between rows increased, because of the use of tractors, so also the distance between the plants in the row increased. As a consequence the total number of plants, as well as of fixed bines, per ha decreased. With the widest quadratic spacing used in Czechoslovakia, i.e. 160 × 160 cm, the number decreased to 3900 plants and 7800 bines per 1 ha (see Table 27).

It was then found that even 160 cm between the rows was not suitable for tractors of normal width. Therefore, the quadratic spacing was replaced by a rectangular one of 260 × 110 cm. This spacing is biologically less suitable for the plants because it departs from the ideal circle, but it had to be introduced in order to allow for mechanical cultivations.

New knowledge about the most suitable number of bines trained from one plant made it possible to change from the two-bine system to one using four bines per plant. It has been shown experimentally, that one plant can safely give up to 6 fully grown bines depending on the age of the plant and the locality. Therefore the optimum of fixed bines per 1 ha became an important element in the planning of the hop plantation. As is clear from Table 27 (page 111) the required planting density was achieved by this new rectangular pattern, using wide spacing and the four-bine fixing system.

The planting of hop gardens with wide spacing formed the basis of a new arrangement of the stand at the start of the 1960s. Spacing of 260 × 110 cm was introduced on flat land, but with the possibility of the equipment slipping on sloping land the spacing at 280 × 110 was introduced there. These spacings were introduced to accommodate wheeled tractors and their equipment, designed as unified packages for hop-garden work. The spaces were such as to allow the implements to pass through without damaging the plants. In this system, the number of fixed bines was approximately 14 thousand per ha. This number was shown experimentally to be the most suitable with regard to achieving the highest yield whilst preserving the required level of illumination in the stand. Thus, the distance of 100 cm between plants in the row was established. The total number of plants per ha had decreased in comparison with the traditional narrow spacing, but after the introduction of the four-bine procedure the total number of fixed bines per ha increased more than 1.5times, and the yield consequently increased by an average of 25 per cent. The introduction of wide spacing changed the whole of hop production technology.

The results obtained by mechanization means and the efficiency of the equipment increased during the 1970s due to an increasing rate of introduction of mechanization. The expected development of hop processing machines foresees the usage of tractors and other equipment of greater efficiency. Some such equipment has already come into operation. Such development also involves greater width and height of the equipment. Therefore new hop gardens in all hop-growing areas of Czechoslovakia have, since 1976 been planted with unified spacing of 300 × 100 cm. Under growing conditions regarded as good or excellent, this means the fastening of 4 bines to 2 poles hanging in a V-shape. Under poorer growing conditions or when there are gaps in the stand it is necessary to use the three-bine system so as to maintain the required light and microclimate conditions in the stand and to achieve the required number of bines (14 000) per hectare. With the wider spacing of the rows, increased from the original 280 to 300 cm, the percentage of plants damaged by cultivation machinery is reduced. At the same time the number of passes made by the tractor, per unit area of crop is reduced by 13 per cent. The introduction of the 300×100 cm spacing also provides a reduction in the capital expenditure for the establishment of the hop-garden.

Cultivation of the soil in hop gardens

The cultivation of the soil in hop gardens has undergone essential changes during the last hundred years. With manual cultivation the soil was dug by hand and a generally flat surface was preserved. Horse-powered cultivation produced a ridged surface well suited to the type of cultivation necessary with narrow spacing. Tractor-powered cultivation in widely spaced rows has returned to a flat surface. Recently, various combinations of flat and ridged cultivations have been examined in hop gardens.

The shaping of hop plants

The overall shape of the hop plant was fully determined by the time of manual cultivation. The correlation between the above-ground and the underground parts of the hop plant as well as between organic systems and between individual organs was usually controlled by the cut.

The structure of the underground system and its various parts is affected by decapitation, cut and pollarding of the plant. Among these procedures, decapitation and the cut have been successfully mechanized and pollarding, usually called the second cut, is now omitted. Decapitation is needed for the modern method of mechanized harvesting. The cut provides the required vernal retardation of vegetative growth and this delay improves the structure of the above-ground part of well-grown plants as well as the quantity and quality of cones.

Cultural interventions during the growing season modify the above-ground part of the hop plant and at the same time increase the number of cones and encourage homogeneous ripening. The fixing of bines has a basic function in shaping the framework and thus controlling the form and structure of the aboveground parts of the plant. The shoot apex of bines is not nowadays shortened as it was in the past because this procedure is replaced by the removal of apical dominance as the apex is bent over the top of the hop garden structure. Subsequent operations modify the length and branching of shoots. The shortening of lower fertile shoots above the second internode causes them to branch more and thus contributes to the production of a larger number of cones. The total removal of infertile shoots aids the better nourishment and better growth of fertile shoots and the cones they bear. Both the interventions involving shoots have been neglected by advances in the technology of production. Recently it has been investigated whether it is possible to replace some interventions, by the use of chemicals.

The technology of large scale production of the hop involves the basic arrangement of the stand, the cultivation of the soil and the shaping and treatment of the plants in the stand.

USE OF CHEMICALS

Chemicals now constitute a necessary part of the production of hop. The changed spacing to extend the distance between rows up to 3 metres made it easier to apply chemicals to the soil surface. Such treatments are mostly used to apply plant nutrients to control certain growth processes by the use of growth regulators, and to protect the stands against weeds, pests and diseases.

Nutrition and manuring

Nutrition of the hop plant

Logically, the manuring of hop stands requires a good knowledge of the typical nutrition of the hop plant. The hop plant obtains its required nutrients – macroelements as well as microelements (except for carbon which is obtained via carbon dioxide) – from aqueous solutions, particularly those in the soil. The best for the nutrition of the hop is a solution with a physiologically balanced content of nutrients. Such an equilibrium is, however, very labile and varies with changes in soil conditions and also due to the physiological activity of the plant. Plants take up from the soil amounts of particular elements, which they require for their metabolism at a given time. Under soil conditions the elements occur in different forms of bonding from which the nutrients can usually be released. Some elements however form resistant compounds in the soil wherefrom it is very difficult if not impossible to release the nutrients for the plants.

Thus, the nutrition of the hop plant can be hindered either by an unbalanced soil solution or by a strong bond of the element in the soil. In the latter case the element is locked in the soil and its uptake by plants is more or less limited. It has been shown, for example that an excessive reserve of the phosphorus in the soil limits the uptake of zinc and causes the so-called curl disease of hop. An excessive content of potassium in the soil limits the uptake of other cations such as magnesium and causes a yellowing of the leaves, especially when there is a lack of water. Otherwise, potassium favourably affects the water economy of the plant by promoting the uptake of water by the roots and by lowering the rate of transpiration. An excess of calcium lowers the uptake of magnesium, potassium and iron and induces chlorosis. Important symptoms of an insufficient uptake of certain macroelements are shown in the following key.

Key for the determination of nutrient deficiency in hop, according to BUREŠ (1975):
1. a) The symptoms appear first on lower leaves and spread upwards . . . . see 2.
1. b) The symptoms appear first on the young parts of the plant. The vegetative apex yellows and decays. The apical leaves become convex, turn yellow and decay. The disorder proceeds downwards, leaves, roll up, they become brown and necrotic from the edges, and decay: deficient in calcium.
2. a) Leaves are smaller, and are dark green in colour. Later they roll inwards, they become light green but develop local brown necroses. These patches may also be orange, red or bronze. Flowering is reduced. At maturity, the cones are not fully closed and the tips of their bracteoles become brown: deficient in phosphorus.
2. b) Leaves are light green to yellow . . . . . see 3
3. a) Shoots are short, whole plants are light green, leaves are smaller and their petioles become red. Later the leaves become bright yellow, with red veins and the leaf surface develops a reddish shade. In later stages the leaves are shed: deficient in nitrogen.
3. b) The shoots are longer than normal . . . . . see 4
4. a) Leaves are of a lighter shade than normal. Old leaves are pale yellow with darker veins; eventually brown spots appear in the spaces between the veins. A 2 – 3 mm wide band around the edge of the leaf becomes dried: deficient in potassium
4. b) Tissue round the main veins of older leaves on the bines becomes yellow after the

second half of July up to the harvest but nearer to the main veins is a border of green. The symptoms proceed from older to young leaves. Yellow leaves become brown. In warm dry weather the leaf tissue can become grey and become boat-shaped as the edges roll upwards: deficient in magnesium.

These deficiencies of various nutrients can occur under conditions either of insufficient or very intensive manuring. Therefore it is important to control and balance the nutrient levels in the soil as well as in the plants. A suitable content of phosphorus and potassium in the soil is shown in Table 28. The limiting values for levels of phosphorus and potassium in heavy soil are given in Table 29. These values are important in determining the amounts of phosphatic and potassic fertilizers to be applied to the soil.

TABLE 28   **Satisfactory amounts of phosphorus and potassium in 100 g of soil**

Type of soil	Reserve of phosphorus (mg)	Reserve of potassium (mg)
Light	13	42
Medium	15	42
Heavy	18	50

TABLE 29   **Evaluation of phosphorus and potassium contained in heavy soil**

Nutrient content	P (mg) in 1000 g of soil (ppm)		K (mg) in 1000 g of soil (ppm)	
	topsoil	subsoil	topsoil	subsoil
Low	less than 110	less than 70	less than 270	less than 170
Medium	110 – 130	70 – 110	270 – 330	170 – 250
High	over 150 – 180	over 110	330 – 380	over 250
Excessive	over 180	–	over 380	–

The total requirement for nutrients has three main components. First is the need for those nutrients in the biomass whose quantity decides whether the required yield of cones will be achieved. In addition there is a requirement for certain nutrients which are not effective during the growing season, but are needed for nutrient reserves.

The total requirement for nitrogen, phosphorus, potassium and magnesium is based on their content in the biomass of a well-grown hop plant. In a mature plant it is equal to the content of these nutrients in cones and in other above-ground organs (except cones).

TABLE 30   **Elements in the above-ground parts of the hop plant**

Date	Total weight of dry matter (g)	Amount of each element (g)					
		sulphur	phosphorus	calcium	potassium	magnesium	nitrogen
14. 6.	122.89	0.278	0.402	2.976	2.546	0.339	4.422
28. 6.	218.86	0.481	0.917	4.749	6.286	0.652	7.419
12. 7.	419.29	0.767	1.359	9.765	11.451	1.086	12.034
26. 7.	606.66	1.232	2.051	14.554	11.723	1.723	14.621
9. 8.	675.31	1.283	2.411	17.160	14.553	1.952	17.355
23. 8.	802.09	1.733	3.072	21.223	18.600	2.422	17.000

Examination of a yield of 1.5 tonnes of dried cones and 3.85 tonnes of plant tissue yielded 113 kg of nitrogen, 20 kg of phosphorus, 124 kg of potassium, 140 kg of calcium and 16 kg of magnesium. The level of nutrient availability depends on the fertilizer used. It is 50–80 per cent for nitrogen, approximately 30 per cent for phosphorus and 40–60 per cent for potassium.

TABLE 31 **Ratio of elements in the above-ground parts of the hop plant**
[Ratio of elements to phosphorus (P = 1)]

Date	Nitrogen	Potassium	Magnesium	Calcium	Sulphur
14. 6.	11.00	6.33	0.84	7.40	0.69
28. 6.	8.09	6.86	0.71	5.18	0.53
12. 7.	8.86	8.43	0.80	7.19	0.56
26. 7.	7.13	5.51	0.84	7.10	0.60
9. 8.	7.20	6.04	0.81	7.12	0.53
23. 8.	5.53	6.06	0.79	6.91	0.56

The required amount of nitrogenous fertilizer to be applied is assessed from its amount in the harvested decapitation product, but for potassium, phosphorus and magnesium it is assessed according to their reserve in the soil. The ratio nutrient uptake to the amount supplied is affected by the use of inefficient nutrients and by the relationship between reserves nutrient and the basic need of the plant. Content and ratio of the main elements shown in Tables 30 and 31 was found by RYBÁČEK (1976) and is for a two-year average.

Manuring of hop

The manuring system usually has two parts:
1. manuring of the soil during the period of vegetative rest;
2. manuring of the plants during the growing season.
During the period of vegetative rest manuring is with farmyard manure which contains reserves of phosphorus and potassium sufficient for two years.

Because the quantity of manure produced each year by the animals on other parts of the hop farms is sufficient for only about one third of the hop gardens, a manuring system was developed with a three-year cycle. This system requires the formation of three groups of hop gardens each of approximately the same area (but they need not belong to the same administrative units). For the system to work it is necessary to analyse the soil of intervals at three years to determine the pH of the soil and the levels of phosphorus, potassium, magnesium and humus.

The yield obtained in previous years is important in determining the required level of manuring, according to the balance method. If the theoretical yield is greater than that actually obtained, then the theoretical yield is used to calculate the basic dose of nutrients and vice versa.

Specification for establishing the required amount of manure, and its application in the hop garden

1. I n p u t   d a t a (to avoid bias the identity of the garden is not revealed in the specification):
   a) type of soil;
   b) planned yield or the real yield (if it is higher than the planned one) in kg dry hop per hectare;
   c) content of phosphorus in mg per kg of the soil;
   d) content of potassium in mg per kg of the soil;

e) content of magnesium in mg per kg of the soil;

f) pH (exchangeable) of the soil.

2. Established amount of farmyard manuring per ha of hop garden: light soil . . . 70 tonnes manuring, medium soil – 55 tonnes, heavy soil – 40 tonnes.

3. Established basic dose (in kg) of the individual elements as indicated from previouse yields:

nitrogen in kg per ha = yield of dry hop in kg per ha × 0.1;

phosphorus in kg per ha = nitrogen in kg per ha × 0.44;

potassium in kg per ha = nitrogen;

magnesium in kg per ha = nitrogen × 0.3.

4. Basic amounts of nutrients are modified according to the results of soil analyses –

a) basic amount of nitrogen is not modified;

b) basic amount of phosphorus is modified as shown in Table 32 a;

c) basic amount of potassium is modified as shown in Table 32 b;

d) basic amount of magnesium is modified as shown in Table 32 c.

5. Manuring system:

a) Farmyard manuring in an amount based on the soil type is made in the first year during the autumn. If this is not possible, then compost is applied in the spring.

b) Manuring by artificial fertilizers –

b a) first year –

basic amount of nitrogen (with ammonium sulphate as the fertilizer of choice);

a double modified dose of phosphorus (best is calcium superphosphate);

basic modified dose of magnesium (as Kieserite);

dose and method of manuring with potassium is determined from the reserves of magnesium in the soil –

variant No 1 – reserve of magnesium less than 100 mg per kg soil, apply basic modified dose of potassium;

variant No 2 – reserve of magnesium higher than 100 mg per kg soil, use doubled modified dose of potassium;

time of manuring – autumn before farmyard manuring ploughed in;

b b) second year –

variant No 1 – basic application of nitrogen (ammonium sulphate);

                – basic modified dose of potassium (potassium sulphate);

                – basic modified dose of magnesium (Kieserite);

variant No 2 – basic dose of nitrogen;

                – basic modified dose of magnesium;

time of manuring – spring, fertilizers applied as preparation is made for cutting;

b c) third year –

liming – in the autumn of the third year a maintenance dose of 650 kg Ca per ha as calcium carbonate ($CaCO_3$) is spread on the soil, if according to soil analysis, more lime is required, then the requisite amount will be applied;

manuring with artificial fertilizers –

– basic dose of nitrogen;

– basic modified dose of phosphorus;

– basic modified dose of potassium;

– basic modified dose of magnesium;

time of application – fertilizers are usually applied before the preparation for cutting in the spring.

6. Selection of fertilizers – principles:

a) it is necessary to follow the principle of compatibility;

b) potassium fertilizers, as chlorides can be applied only during the autumn, in the spring it is necessary to use potassium fertilizers as sulphates;

c) nitrogen fertilizer must be selected in a slow-release form;

d) if the value of exchangeable pH exceeds 7, then physiologically acid or neutral fertilizers should be selected.

TABLE 32a   Soil reserves of phosphorus and its manurial requirements

Quality of soil reserves	Phosphorus in top soil (mg per kg)	Increase over normal requirement of phosphorus (kg per ha for 3 years)
Low	less than 100	45
Medium	101 – 150	30
Height	151 – 180	0

If there is more than 240 mg phosphorus per kg of topsoil the phosphorus application can be reduced to half (this concerns the basic dose as determined in point 3).

TABLE 32b   Soil reserves of potassium and its manurial requirements

Quality of soil reserves	Potassium in top soil (mg per kg)	Increase over normal requirement of potassium (kg per ha for 3 years)
Low	less than 270	45
Medium	271 – 330	30
Height	331 – 380	0

If there is more than 400 mg potassium per kg of soil the potassium application can be reduced to half (this concerns the basic dose as determined in point 3).

TABLE 32c   Soil reserves of magnesium and its manurial requirements

Quality of soil reserves	Magnesium in top soil (mg per kg)	Increase over normal requirement of magnesium (kg per ha for 3 years)
Low	less than 60	45
Medium	61 – 100	30
Height	101 – 130	0

If there is more than 160 mg magnesium per kg of soil the magnesium application can be reduced to half (this concerns the basic dose as determined in point 3).

Fertilizers used for hop manuring

Organic fertilizers

Farmyard manuring is a basic and proven fertilizer for the manuring of hops. It contains 0.5 per cent of nitrogen, 0.09 per cent of phosphorus, 0.5 per cent of potassium, 0.3 per cent of calcium and around 20 per cent of organic substances as well as various microelements depending on its origin. The amount to be applied is governed by the type of

128

1. Phosphorus deficiency.

2. Potassium deficiency.

3. Calcium deficiency.

4. Chlorosis of leaves due to deficiency in magnesium and iron.

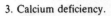

PLATE I.
Symptoms of deficiency in several nutrients.

PLATE II.
Damages to hop plants.

1. Herbicide damage.

2. Damage caused by sulphur dioxide.

3. Leaves burned by preparation "Reglone" (two stages).

PLATE III.
Damages to hop cones.

1. Damages to cones due to
wind, cones knocked off.

2. Cones dried out (physiological disorder).

PLATE IV.
Downy mildew of hop, caused
by *Peronospora humuli*.

1. Leaves showing different
levels of disease severity.

3. Healthy leaves and leaves attacked by
Peronospora humuli.

2. Hop shoots attacked by *Peronospora
humuli*.

PLATE V.
Downy mildew – *Peronospora humuli* and powdery mildew – *Sphaerotheca humuli.*

1. Healthy cone and different stages of damages by *Peronospora humuli.*

3. Different stages of the attack by powdery mildew.

2. Powdery mildew.

PLATE VI.
Hop aphids.

1. Overall effect of in festation on the stand.

2. Cones damaged by hop aphid.

3. Larva of drone fly devouring aphids on obverse of hop leaf.

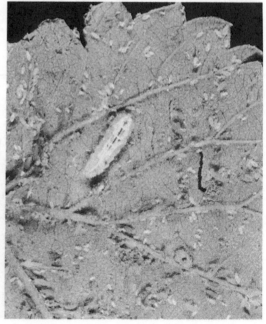

PLATE VII.
Pests.

1. Leaf damaged by spider
mite

2. Otiorrhynchid beetle.

PLATE VIII.
Hop garden and its product.

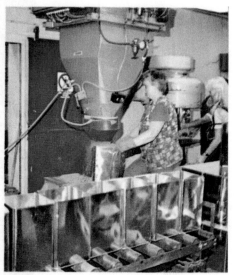

1. Training of deviated vegetative apices.

2. Filling of tins with granulated product (Hop Growing Company, Žatec).

3. Harvest centre of pilot farm of Research and Breeding Institute for hop growing in Stekník (district Žatec).

soil – heavy soils are satisfied with 40 tonnes per ha, medium soils with 55 tonnes per ha and light soils with 70 tonnes per ha. It is usually applied during the autumn.

C o m p o s t is the second most important organic fertilizer. A good compost contains over 50 per cent of organic substances in its dry matter and has more than 2 per cent of nitrogen, 0.6 per cent of phosphorus, 0.2 per cent of potassium and over 4.5 per cent of calcium and magnesium. It needs to be of loose consistency, it should have the scent of garden soil, it should be well decayed and its original components should be not identifiable. It can be used instead of farmyard manuring during the spring.

L i q u i d   m a n u r e is important for its content of auxins, nitrogen and potassium. *It is applied in two different ways:*

a) dry conservation – the slurry is poured over either compost or farmyard manure;

b) direct manuring of plants (the liquid manure is usually applied in the growing season at 150 hl with added superphosphate plus 300 hl water per ha, the latest time for its application is at the beginning of flowering).

G r e e n   m a n u r e has provided good results in Mddle European conditions when seeded 10–14 days before the hop harvest (RYBÁČEK, 1962, ŠKODA, 1976). Crops to be used include white mustard, tancy (*Phacelia*) or field pea. With an expected germination capacity of 94 per cent, the seeding rate for white mustard is 12–15 kg per ha and for tancy (Phacelia) 20–22 kg per ha of hop garden. The soil surface should be harrowed before and after seeding and a narrow gauge seeding machine is needed. After the application of green manuring herbicides of the triazine series and substituted urea are not used.

Where herbicides are used before the seeding it is necessary to test seed germination in a sample of soil with the green manure and in a control, e.g. sand, under the same watering and temperature conditions. If seedlings do not appear, it is better to give up the green manure. The best period for ploughing in the green matter is in the autumn, before the first frosts.

C a r b o n a c e o u s   m a n u r e s are produced in a patented process from fine, floating particles of brown coal. They contain 9.5 per cent water, 40–50 per cent of organic matter, 3 per cent nitrogen, 0.88 per cent phosphorus, 2.5 per cent potassium, 3.55 per cent calcium and various microelements (boron, zinc, cobalt etc.). This fertilizer is black and granular. Its most suitable time of application is during the autumn or early spring. It is used mostly with light and medium soils. The application varies from 4 to 7 tonnes per ha.

Artificial fertilizers

Fertilizers consisting of one component

N i t r o g e n o u s   a n d   p h o s p h a t i c   f e r t i l i z e r s are shown in Tables 33 and 34.

The p o t a s s i c   f e r t i l i z e r s (see Table 35) are preferred, which contain magnesium. Fertilizers in the form of sulphates are used in the spring on soils deficient in calcium.

M a g n e s i t e   f e r t i l i z e r is represented by Kieserite which contains 14–16 per cent magnesium, as sulphate and 20 per cent sulphur. It works quickly and has a weak acidic physiological reaction. Magnesium is also present in c a l c i u m   f e r t i l i z e r s .

Fertilizers consisting of more than one component

The following composite fertilizers are used for the manuring of hops:

C e r e r i t e with 11 per cent nitrogen, 4 per cent phosphorus, 12 per cent potassium and 3 per cent magnesium (potassium is in sulphate form, the fertilizer also contains boron, molybdenum, zinc and copper).

R u s t i c a with 12 per cent nitrogen, 5.23 per cent phosphorus, 14.11 per cent potassium,

TABLE 33 Nitrogenous fertilizers

Fertilizer	Nitrogen content (%)	Other constituents	Speed of effect	Physio-logical reaction
Ammonium sulphate	21	58 % SO₃	slow	acidic
Calcium cyanamide	16 – 18	39 % Ca	slow	alkaline
Urea	46		slow	mostly neutral
Calcium nitrate	15	22.5 % Ca	quick	alkaline
Ammonium nitrate	33		slow and quick	mostly neutral
Ammonium nitrate plus calcium	25	11 % Ca	slow and quick	alkaline
Ammonium nitrate plus calcium	30	5 % Ca	slow and quick	alkaline

TABLE 34 Phosphatic fertilizers

Fertilizer	Phosphorus content (%)	Other consti-tuents	Speed of effect	Physiological reaction
Superphosphates				
Cola	8	CaSO₄, Mn	quick	neutral
Afrika	6	Cu, Zn	quick	neutral
Refos[1]	7.9	CaSO₄	quick	neutral
Granular[2]	7	CaSO₄	quick	neutral
Thomas powder[3]				
From Kladno	7[3]	33 % Ca plus trace elements	slow	alkaline
From Vítkovice	5.3[3]	2 % Mg	slow	alkaline

Notes:
1) 5 % water soluble phosphorus,
2) 6.6 % water soluble phosphorus,
3) soluble phosphorus citrate.

TABLE 35 Potassic fertilizers

Fertilizer	Potassium content (%)	Other con-stituents	Speed of effect	Physiological reaction
Potassic salt	32 – 35	46 % Cl	quick	acidic
Potassic salt	40 – 44	47 % Cl	quick	acidic
Kamex	32 – 35	43 % Cl 2.4 % Mg	quick	acidic
Potassium sulphate	40 – 44	17 % S	quick	acidic

TABLE 36   Calcareous fertilizers

Fertilizer	Calcium content (%)	Other constituents	Speed of effect
Quicklime	63 – 71	up to 6 % Mg	quick
Ground limestone	40 – 44	up to 3 % Mg	slow
Saturation dregs from sugar factory	14 – 17	0.17 % P, 1.2 % Mg, 0.2 % N	slow
Basic slag	23 – 31	1.8–5.4 % Mg, P and S plus traces of other elemets	slow

1.2 per cent magnesium and traces of boron, manganese, copper, zinc and cobalt (half of the phosphorus is insoluble, potassium is in the form of the sulphate).

N P K 2 A with 11 per cent nitrogen, 5 per cent phosphorus, 12 per cent potassium, 1 per cent magnesium (nitrogen is as ammonia, phosphorus is 2/5 water soluble and potassium is as chloride).

N P K S y n f e r t a is similar to NPK 2A, but without magnesium.

Factors controlling the rate of application of fertilizers

The application of artificial fertilizers is based on a theorical knowledge of the principles controlling the nutrition of the plant, together with practical experience of the effect of ecological factors on nutrient uptake and utilization as they affect yield. This knowledge must be applied to a rational manuring system. The aim is to achieve high yields of top-quality hops under economically favourable conditions and without harmful effects on the environment.

Fig. 55. Deep insertion of liquid fertilizers with subsoil plough.

Fig. 56. Surface application of liquid fertilizers.

131

The efficiency of artificial fertilizers is relatively low, as the plants utilize only a proportion of the applied compounds. The highest utilization rate, 50–80 per cent, is that of nitrogen, phosphorus reaches only 30 per cent and potassium is between 40 and 60 per cent. The reasons for this are to be found in the chemical composition of the fertilizers as well as in their reaction with the soil. Phosphorus, for example, produces compounds with Ca^{++}, Al^{+++} and Fe^{+++} whose solubility is very low. Losses of nitrogen, potassium as well as calcium are washed away by gravitation water. The rate of loss of nutrients by leaching depends on their form in the fertilizer, the type of soil, the soil reaction (pH), climatic factors and agrochemical measures.

The influence of *the form of nutrient* can be demonstrated with, for example, nitrogenous fertilizers containing the nitrogen as the NO_3^- or NH_4^+ ions, or in organic form. The greatest losses occur with the NO_3^- form because it is not physically bound in the soil and is consequently leached away (70 per cent or even more). The NH_4^+ form is bonded in the soil to the sorption complex and therefore its losses are rather less (50 per cent or more). When nitrogen is in the organic form (e.g. $CaCN_2$, $CO(NH_2)_2$), its efficiency depends on the rate of its mineralizing in the soil. If correctly used, the losses do not exceed 20 per cent. Very important here is the mechanical property of the soil, its pH value and the activity of microorganisms in the soil. Fertilizers with the NH_4^+ ion can suffer significant losses to the atmosphere if calcareous fertilizers are not used correctly.

The influence of *the type of soil* depends on its composition: soils with a high content of argillaceous particles have a marked ability to bind any applied chemical, light soils, on the contrary, have a much weaker ability to bind. This property of binding is also controlled by the soil content of biological material and its biological activity. The binding of nutrients in the soil increases as the content of permanent humus increases.

When *the pH of the soil* is high, particularly in the presence of a high content of Ca^{++} ions the losses of nitrogen are greater. Lower pH value can favour the movement of the NO^3 ion in the soil.

Among the climatic factors, it is the amount of precipitation, the temperature and the movement of the air which affects the circulation of nutrients in the soil and consequently affects their losses. Precipitation controls the losses of nutrients due to leaching in proportion to rate of water. It has been found that 150 mm precipitation water leaches the soil profile down to 80 cm. The movement of the Ca^{++} ion produces so called "calcareous horizon" on the impermeable layers of the subsoil. The greatest losses of nutrients due to leaching occur on light soils. Higher temperatures cause losses of nitrogen by its volatilization and by affecting denitrification.

Those agrotechnical measures which improve the root system contribute to the better utilization of fertilizers and limit the losses of nutrients.

Losses of nutrients can be reduced by the following methods:
1. improvement of soil properties;
2. utilization of inhibitors of nitrification;
3. improvement of the properties of artificial fertilizers;
4. using the correct method of application of fertilizers.

It is important to achieve the full utilization of the available artificial fertilizers to apply them only when required and to apply the correct dosage.

Biologically active substances – growth – controllers

The growth of hop can be regulated by different biologically active substances. These include β-indolyl-acetic acid and β-indolyl-γ-butyric acid. POKORNÝ (1972) obtained good results in *the rooting of cuttings* with weak solutions of the β-indolyl-acetic acid (200 mg per

1 water soak for 24 hours) and the β-indolyl-γ-butyric acid (200 mg per 1 water for 24 hours) or their mixture (100 mg of each in 1 l water for 24 hours). Steeping in the IAA solution promoted particularly vigorous rooting.

The growth of hop can be also regulated in the spring by the use of certain herbicides.

The spring growth of hop can be retarded and the spring cut avoided with the help of different preparations like Reglone (with diquat in solution 200 g per 1 as active substance), Prevenol (with chlorprophan in solution 392 g per 1 as active substance) or Prefix (with chlorthiamid as active substance). A hop-garden sprayed in the spring but before sprouting with Prevenol (0.5 kg of preparate per 1 ha) showed an approximately 3 week delay in sprouting (SACHL, 1969).

WEISER (1977) quoted results obtained with various compounds investigated in the German Democratic Republic (Alar, gibberellic acid, CCC, BMH, Ethrel, Malzid). He particularly studied the effect of growth regulators on *the development of cones* and the consequent *higher yields* as well as *the levels of compounds contained in them*. He also examined compounds for their effect on *shortening hop bines* and *improving their climbing* because this would make the fixing of hop shoots easier and could eventually allow a reduction in the height of hop garden structures.

Two sprayings of 0.1 per cent solution of the compound Alar 85 (daminozide = N-dimethylaminosuccinic acid) increased the weight (mass) of cones and their number and ,improved the content of resins.

The tests of the influence on the yield and quality of hop cones of gibberellic acid had different results. Two concentrations were tested (5 mg and 10 mg per 1 l) with different application intervals. The number of cones was increased but their weight was essentially decreased and the content of α-bitter acid was also lower.

In U.S.A. the upper parts of hop plants were sprayed with gibberellic acid at 10 mg per 1 l and the crop of cones increased as well as the content of humulone and lupulone.

The results of experiments in England were just the opposite although plant growth was increased; this increased growth was attributed once to the variety once to the weather (KUTINA, 1988). KARABANOV (1963) mentioned that an application of gibberellic acid was followed by a change to a lighter colour of leaves and the yield of cones increased on old plants but decreased on young plants.

WEISER (1977) obtained unsatisfactory results with CCC = chloromequat chloride (2-chloroethyltrimethylammonium chloride) and with BMH (N-dimethyl-N-2-bromethyl-hyd-razonium bromide). Shortly after the application of a 0.5 per cent solution leaves became yellowed. This change of colour was particularly noticeable after the application of chlorme-quat chloride. Leaves became necrotic and were shed. The expected effect of shortened bines did not appear. Likewise, the application of gibberellic acid was not successful; the expected shortening of internodes and an increased number of cones did not occur.

Ethephon (2-chlorethylphosphonic acid, as Ethrel) was tested at concentrations from 0.05 to 2 per cent and at different time intervals. Concentrations of 0.1 per cent and above retarded the growth of the plant and reduced the climbing activity of bines. The apices of bines were so affected that they became oriented downwards, and the number of flowers was much reduced.

Maleic hydrazide (1.2-dihydropyridazine-3.6-dione) was tested at concentrations of 0.05 per cent to 5 per cent. With concentrations higher than 0.1 per cent, the treated plants reacted with shortened internodes and with the growth of the apices of shoots almost completely inhibited. The production of inflorescences was apparently unaffected.

In Czechoslovakia a multi-component liquid fertilizer Vegaflor is used in hop-growing practice for foliar nutrition of the hop.

Chemical weed control in hop gardens

The use of herbicides for the chemical removal of weed in the plantations, rooted-cutting nurseries and hop gardens makes it possible to reduce the number of cultural operations.

Chemical control of weeds in the productive hop garden

Two seasons are critical with regard to the possible weed infestation of productive hop gardens; the spring period, from cutting to first extra-ploughing, and the summer season from the end of July up to the end of vegetative growth. In principle, therefore the two periods for the application of herbicides are in spring, after the cutting, and in summer after the extra-ploughing.

Weed control using herbicides in the spring

The application of herbicides in the spring aims to remove current annual weeds in that period between cutting and extra-ploughing when weed control is impossible by means of cultivating and ploughing. Moreover, in wet sites and in years with a rainy spring the extra-ploughing is done too late, when the weeds are already well grown and cannot be destroyed by ploughing.

For spring treatment of hop gardens following preparates can be used: Afalon (47 per cent of linuron, at 2–3 kg per 1 hectare), Gesagard 50 (50 per cent of prometryne) and Gesatop 50 (50 per cent of simazine used at 3–4 kg per ha – smaller amounts for hop gardens with light soil and a lower content of humus). The volume of sprayed liquid should not be less than 500 l per

Fig. 57. Spring application of herbicides in the hop garden.

ha. All presently available spraying machinery, with a frame for surface spraying, can be used for this treatment. The working width of the equipment should be modified to fit the spacing of the poles bearing the wirework.

The spraying must provide homogeneous treatment of the hop garden. The jets need to be so arranged that their spray patterns do not overlap, and likewise, there should be no overlap of the sprayer zones as the sprayer passes through the hop garden and turns at the end of the row. It is vital to avoid local overdosing with the consequent damage to plants if the spraying is not correctly controlled when the sprayer stops or starts.

All the compounds mentioned above are able to prevent weed infestation of hop gardens in the period between cutting and extra-ploughing. They also have a residual effect so that the amounts of weed will be reduced even after extra-ploughing. Therefore, it is usually possible to omit the second extra-ploughing and to reduce also the amount of summer cultivation. This spring treatment with herbicides can be applied after the cutting and prior to stringing, and certainly before the sprouting of hop shoots.

Preparations based on simazine (Gesatop 50) are suitable for the humid areas of hop-growing regions or for wet sites and must be applied before weed seed germination.

The compounds Afalon and Gesagard can be applied to weed seedlings, but prior to the sprouting of the first hop shoots. Their efficiency is less dependent on the moisture in the soil, therefore they can provide good results even in relatively dry field sites.

The maximum benefit from herbicides requires the correct preparation of the soil surface; the tilth of the soil must be the finest possible.

Weed control by the use of herbicides in the summer

The treatment of hop gardens with herbicides in summer aims to hinder weed infestation during the summer and autumn. Besides those herbicides usually used for the spring treatment, it is possible to use Gesaprim 50 (50 per cent atrazine) after the extra-ploughing. The application rate for all compounds is 2 kg per ha in 600 l of water.

In practice the summer treatment is usually with Gesaprim 50, because its long residual acting ensures the complete absence of weeds from the time of extra-ploughing through to winter tillage. Weeds take up this compound through both roots and leaves, so its efficiency is not dependent on moisture in the soil or precipitation. The treatment is not applied immediately after extra-ploughing but after the germination of the weed seeds. The effect of the herbicide is hence increased and the result is practically independent of external conditions. This indicates why the summer application is preferred over that in the spring. Hop gardens treated in this way do not require any mechanical cultivations.

For the spring treatment of hop gardens with herbicides any sprayer can be used if it is equipped with a frame specially modified for surface spraying with a working radius up to 3 metres.

Chemical after-cleaning of hop

This treatment applies to the chemical removal of new shoots produced after the extra-ploughing and the defoliation of the lower parts of the bines. Reglone (20 per cent diquat) can be used here. This compound is a non-selective contact herbicide, effective against a large group of plants, mostly dicotyledonous. It can be applied to hops because it does not harm tissues protected by sufficiently thick epidermis. It can be applied at 4 kg of the preparation in 600 l of water per ha after the extra-ploughing, providing that the hop plants have reached a height of at least 5 m. The spraying should be such that the treatment does not reach the fixed shoots but should cover only the lowest floor of leaves. A low pressure, up to 0.3 MPa, should be used and spraying should not be done when there is a wind, because there is a danger that the hop leaves would be damaged by blown spray. It is also essential to flush, thoroughly, the tank of the sprayer after use.

Reglone is not recommended for the extermination of established weeds in a very infested hop garden. In such cases the application rate and the total volume of the fluid would be very large, leading to a much increased risk of damage to the hop plants. Moreover, the economics of treatment in such a situation will be also very disadvantageous.

Chemical control of weeds in nurseries

This type of weed control produces very good results and has provided the base for a new technology for producing plants with a minimum of manual work. The advantages of this chemical method can be used only by those farms which have sufficient experience of the use of herbicides in productive hop gardens. The risk of damage to young plants by herbicides is very high and therefore it is necessary to adhere precisely to the correct application procedures.

Among the available compounds, linuron and simazine can be used in top nurseriers, at a rate of 2 kg of the preparation in 600 l of water per ha.

The compound Gesatop 50 (simazine) is very safe because it has very low mobility in soil and therefore it is particularly suitable for pot nurseries with light soil. The treatment has to be applied before the weed seeds germinate. If the soil is not sufficiently moist, it is necessary to wet the ground, with at least 15 mm of water ten days after treatment.

Afalon (linuron) is suitable for use in top nurseriers with no watering system. If the soil is not wet enough then this compound acts on the weeds through their leaves. As there is a difference in the starting time of growth, between hops and weeds, it is possible to treat plants in the time interval between the germination of weed seeds and the sprouting of hop shoots. The weeds will dry up within a week after treatment and the overall efficiency is satisfactory even when the soil is dry.

Both linuron and simazine can destroy the weed present in the garden in the time interval between sowing and growth. Their application in top nurseriers is safe providing the seeds are at least 10 cm below the surface of soil and that the soil surface is properly levelled after seed sowing.

It is essential to measure exactly the chemicals for the preparation of spraying fluid and to the correct spraying procedure.

Crop protection in the hop garden

Contemporary large-scale hop production in specialized farms, concentrated in strictly limited regions, produces considerable changes in the ecological system and provides conditions for the development, propagation and over-population of various harmful factors. Therefore present-day hop production is unthinkable without an intensive crop protection programme.

A successful protection against hop diseases and pests is based on a good knowledge of the particular harmful factor and on the correct use of chemicals and machines. Also necessary are a knowledge of various laws, instructions, standards and regulations in the use of chemicals and in relation to hygiene and safety of work. This chapter is therefore arranged to provide a survey of the most important pests and diseases of hop together with the main principles of crop protection.

Practical details are regularly described in a hand-book for the protection of plants, published in Czechoslovakia, which is updated every year and in other publications concerning the problems of plant protection.

136

Diseases of hop

The causes of diseases and the reasons for damage to hop plants can be classified into three groups:

a) abiotic factors causing physiological diseases;
b) viruses inducing virus diseases;
c) microorganisms (including *Bacteria* and *Fungi*) causing parasitic diseases.

Physiological diseases

Physiological diseases result from different harmful factors of non-parasitic origin. In stands of hop, such diseases caused for example by poor nutrition and unfavourable meteorological conditions reduce the yield and quality of cones.

P r e m a t u r e   w i l t . This has recently become one of the most important diseases of hop. First symptoms appear on the plants as early as 10th – 15th June when the bracts and the apical parts of the shoots down to an internode become brown. Soon the whole apex of the shoot dries and a boundary zone appears at the next node. The same symptoms also appear on the secondary shoots. Those shoots whose apices continue to grow remain infertile. Where the apices later become dry, inflorescences and cones appear, but they too become successively dried. As the inflorescence dries the rachis becomes brown, a callus layer is formed and finally the dried inflorescence becomes brown and falls off.

Drying cones are usually small and open and they are lighter in colour than are healthy ones.

The rachis becomes brown and finally the bracts too become brown. A dried cone can be easily removed from the plant, unlike the cones dried as a result of attack by Peronospora or excess of phosphorus.

This disease is significantly more frequent on plants trained as reverse conical than on cylindrical plants. The drying symptom is usually accompanied, particularly in over-dense stands, with the leaves becoming yellow step by step according to their level on the bine.

In Middle European conditions, this disease is reduced by balanced manuring applied according to the results of soil analysis and with liming every third year. If the soil is boron deficient this microelement is supplied. New hop gardens must have a spacing which provide optimum lighting conditions and only two pairs of plants can be fixed.

C u r l   i n   h o p s . This disease is known in the hop-growing regions of Czechoslovakia, the Soviet Union, Poland and both German states. In the past in was assumed to be a virus disease closely related to nettlehead, because both have certain similar symptoms. However, it was never proven to be caused by an infectious agent, and its association with deficiency of zinc in the soil was independently discovered in Czechoslovakia, the German Democratic Republic and the Federal Republic of Germany. Experiments on nutrition showed that if the curled plants are regularly supplied with zinc during the developing period of the vegetative organs, plant growth is quite normal, and moreover, that the symptoms of the disease disappear, together with all changes in the vital processes and biochemistry of the plants caused by the disease. Thus, applications of zinc compounds can be used to heal curled plants and to increase the yield and quality of cones.

The symptoms of curl vary considerably and fluctuate in particular years from a latent indisposition right up to the drying and death of the whole plant. The intensity of symptoms varies according to the environmental conditions. Plants with weak symptoms have long shoots with tiny leaves and thin inflorescences. The leaves are somewhat deflected upwards and are boat-shaped, their crenation is sharp and the lamina are deeply lobed. The symptoms usually appear at the end of June and during July. Heavily attacked plants are dwarfed, greenish-yellow, their shoots are short and grow upwards close to the bine. The leaves on the bines are a lighter colour, even translucent yellow between the veins and the blades become dried, successively outwards in long strips. Leaves are fragile, they break easily and the whole plant rattles if moved. The cones on attacked plants are small, they remain open, and their quality is very low. With a high proportion of curled plants the productivity of the stand is reduced by 50 to 70 per cent.

The economic importance of curl continues to be great. Investigations showed that in the period from 1965 to 1967 this disease affected the yield and quality of cones on approximately 40 per cent of the total hop-growing area in Czechoslovakia. After 1967 when curl started to be treated with zinc compounds its harmfulness quickly decreased.

To heal a stand of curled plants, zinc sulphate ($ZnSO_4.1H_2O$) is applied through a sprayer to leaves or to the soil. The spray is used 0.1 to 0.15 per cent. The first treatment must be applied immediately after fixing of the plant. In order to heal heavily attacked plants four treatments are needed in the period between fixing and the production of cones. It is not necessary to spray zinc separately; it can be applied with sprays against Peronospora. If the zinc compound is applied only to the leaves, the healing of the plants is only temporary and has to be repeated annually. If zinc sulphate is applied to the soil, then the healing is more durable, and. according to the amount applied, may last for up to 5 years. It is recommended to apply zinc sulphate combined with ammonium sulphate at 100 kg $ZnSO_4$ plus 250 kg $(NH_4)_2SO_4$ per ha. This soil application is made either in the autumn or in the spring and applied to ploughed hop garden in rows around the plants. The mixed preparation must be used immediately, it cannot be stored.

D a m a g e c a u s e d b y c o l d a n d f r o s t . Low temperatures retard the growth and development of the hop plant. In Middle Europe cold weather and late spring frosts occurring as late as May are the most dangerous. Cold weather causes the leaves in the upper parts of the plants to become yellow (so called icteric discoloration). The leaves of the lower parts of plants, in contrast, remain green. This is a temporary phenomenon and disappears once the temperature rises. If temperatures fall below freezing point, hop rapidly becomes rigid when the tissues become brittle and tend to fracture. If the temperature falls to −1 or −3 °C, the bine leaves roll inwards forming a fist shape and their edges become black and dried. The plants revive with a slow rise in temperature. Severe frost so damages the apices and leaves that they fail to recover.

S u n s t r o k e . Sunstroke appears after training, usually in June and July. But does not recur during subsequent vegetative growth. It usually develops on light soils when sudden dry hot and sunny weather follows a humid and cold period. Thus the main reason is a sudden high temperature with intensive solar radiation. The microclimate is indirectly associated with sunstroke in relation to the height of damage on leaves. Most damage usually occurs around the middle of the plant, i.e. at a height of about 2–4 m above soil level. Here the extremes of temperature tend to be concentrated and it is also here that radiation of the sun is focused during daylight.

In leaves affected by sunstroke strips between the main veins change colour and suddenly become dark brown. The affected leaves roll upwards, become dried and are likely to be shed. Sunstroke can be confounded with curl but this usually appears later.

Loss of a small proportion of leaves will usually be made up during next growth period. A quicker regeneration will be attained if greater care is taken of cultivation, and with suitable manuring during the next growing period.

K n o c k e d c o n e s . Cones often suffer damage in hop gardens exposed to the wind and in the border rows. The damage results from friction between cones and leaves, shoots and bine which results in small brown necroses at the tips of the bracts. If the cones are more buffeted, more of the bracts become brown and dried, or sometimes the whole cone changes. Inflorescence can be damaged by the wind and such inflorescences later become dry and fall off. Strong winds often break off whole fruiting branches. The appearance of knocked cones suffers greatly and their commercial value is reduced. Once they are damaged the appearance of knocked cones cannot be changed.

This type of damage can be avoided by the careful selection of suitable sites for hop gardens and by the concentration of gardens into large blocks so that the proportion of plants exposed to the wind is as small as possible.

H a i l . Hop plants are very sensitive to damage by hail. The vegetative apices are normally the most damaged parts. The leaves, particularly the bine leaves, may be perforated or even totally destroyed. Hail before harvest may cause a change in colour and a certain amount of drying and decay of cones; more severe hail may destroy the cones and may even destroy the whole above-ground part of the plant.

The amount of damage as well as the scope of protective measures depend on the season. If the shoots are damaged at the beginning of the growing season (up to the first half of June) it is possible to remove these shoots and to train new shoots. However, since it is essential that the cones are all at the same stage of maturity at harvest the whole hop garden should be treated similarly. If the plants are damaged when they have reached a considerable height the growing cones would also be damaged. In such a case the bine is decapitated above the highest undamaged node and the best side branch is trained as a new apex, all branches below this new apex being removed. Plants damaged by hail regenerate quickly if a rapidly-acting potassium and nitrate fertilizer is applied. Fertilizer is given even when plants are damaged during the second half of growing season (July, August) in order to aid recovery, to encourage the transfer of some of the reserve substances back to rootstock, and to hinder their breakdown.

D a m a g e d u e t o i n d u s t r i a l p o l l u t a n t s . Gaseous fumes are more harmful to the hop than is the particulate flue ash, because they are damaging at very low concentrations and can spread out to considerable distances from their sources. Damage caused by sulphur dioxide has been observed at the boundaries of hop-growing regions. It is usually seen on leaves and cones because they are particularly sensitive. Low levels of damage to hop occur as the long-term effect of sulphur dioxide at a concentration as low as 0.15 mg per 1 m^3 of air. A concentration 0.42 mg per 1 cubic meter of air will cause damage to leaves and cones after 4–5 h of exposure depending on the weather conditions.

At the beginning small, clearly visible spots appear on leaves. At higher concentrations leaves lose turgor, become brown-yellow and dark necrotic spots, limited by the veins appear on leaf blades. Severely damaged leaves roll upwards, dry out and are shed.

First appearance of damage on the cones is seen as minute light brown spots at the tips of the bracteoles. These spots later spread to cover the whole of the tips. Under the effect of high concentrations the cones become brown and dry out. According to the degree of damage the yield may decrease by up to 70 per cent. At the same time the content of α bitter acids decreases and the brewing quality of the cones is reduced. Hops damaged by atmospheric pollutants are reduced to lower categories of quality and do not meet export standards.

Fly ash is deposited on the surface of leaves where it interferes with the access of light and consequently limits the intensity of photosynthesis. It also forms an unsightly layer on moist bracteoles.

The main sources of pollutants are the power stations sited at the bounderies of the hop-growing region of Žatec. The maximum danger of harm is at a distance of approximately 5 to 10 km from these sources. Therefore, the renewal or extension of hop gardens in this zone cannot be planned.

Viroses

The systematic registration of mother hop gardens, producing seed, was introduced in Czechoslovakia more than a quarter of century ago, and is highly effected. But this administrative procedure could not take into account those diseases, whose infectious

Fig. 58. Line pattern.

character was not recognized or for which no objective diagnostic methods were known. Accordingly, as a result of recent knowledge (KŘÍŽ, 1967, SCHMIDT, 1972) the disorder known as *curl,* caused by deficiency of zinc in the soil, is not classified as an infectious disease, nor is infertility because it is a genetic deviation. However, it has been clearly shown that line pattern, split leaf blotch, hop mosaic and nettlehead are caused by viruses.

L i n e   p a t t e r n . This is a disease known in all hop-growing regions and it was described in Czechoslovakia in 1938 by BLATTNÝ. It is characterized by light green or yellowish patterns of circles and stripes occurring irregularly on the leaves at the end of June. There symptoms are normally limited to lower leaves, they do not appear on leaves in the upper parts of the plant. Sometimes these weak symptoms disappear during the growing season. Sometimes the first symptoms appear later, in the second half of July, and are limited to a few leaf layers in only the upper third of the plant. Leaves with symptoms of line pattern are usually somewhat wavy in appearance. The symptoms are clearly apparent only in certain years, often they do not appear at all, or perhaps shortly before harvest.

Another form of this disease is identified by different symptoms, with similar but sharper designs on all leaves which become brown and necrotic. The symptoms of this form of line pattern are always of the same intensity regardless of the climatic conditions and this form of the disease does not continue into the dormant period.

Latent forms of line pattern do not usually affect the growth and development of the plant nor its preproductive ability. But plants with symptoms of necrotic line pattern are always weak; they produce only a small amount of cones which remain dwarfed because the assimilatory apparatus of the leaves is impaired and the brewing quality of the cones is low. Also, in these cases the reproductive ability is reduced by up to 50 per cent.

Line pattern is transmitted by inoculation grafting to indicator hops *Humulus lupulus* subspecies *neomexicanus, Humulus japonicus* and variety 'KAV' (clone 'Aschersleben no.

140

V'). Certain indicator plants can be successfully used for diagnostic purposes especially *Chenopodium quinoa* Wild. and *Cucumis sativus* L., which show typical lesions as early as one week after inoculation.

Studies made in Czechoslovakia and the German Democratic Republic have shown that line pattern can be found on all economically important varieties of hop. This being the case the process of protective intervention is focused on negative selection within the framework of the registration procedure for the mother hop garden. Here, plants with symptoms of necrotic line pattern are marked and removed from the stand.

S p l i t   l e a f   b l o t c h . The first symptoms of this disease appear very early, usually before the training of the hop plants. The leaves on bines demonstrate a typical deformation. Because sections of the venation become necrotic, individual lobes and often whole leaves become deformed and roll inwards. The leaf blade tissue between the main veins is wrinkled and the blade itself becomes perforated. In addition to the necrosis of individual sections of the leaf veins there is another typical symptom, namely longitudinal necroses on the bine, shoots and leaf petioles. The growth of such an infected plant is usually less vigorous, the length of internodes remains short, the vegetative apex loses its climbing ability and falls away from the training wires. Plants come into flower early but poorly. The disease does not have a latent phase. Its symptoms are always clear, independently of external circumstances.

The disease can be transferred by grafting to the indicator variety 'KAV' and by inoculation to the same indicator plants as are used for line pattern. The results of pre-immunity testing on *Nicotiana megalosiphon* (SCHMIDT et al., 1965) show that the virus of split leaf blotch has the same origin as that of line pattern. Up to the present this disease has been identified in Czechoslovak, East German and Polish hop-growing regions. In spite of the fact that the yield of attacked plants is decreased by two thirds, split leaf blotch is not of great economic importance because of its limited occurrence. Its symptoms are clearly visible throughout the growing season so that negative selection provides satisfactory results.

H o p   m o s a i c   (E n g l i s h   m o s a i c) . The causal agent of this disease is *Humulus virus 1* Salmon and was described in England by Salmon in 1923. It was subsequently found in the USA, Australia and in almost all European hop-growing regions. In the USA and UK it is a very serious so-called limiting disease of the hop plant but in Czechoslovakia it occurs only sporadically and is consequently of only limited importance. Its symptoms differ in different varieties of the hop plant. Infection is very apparent in varieties of the "Golding" group but other varieties (e.g. 'Fuggle', 'College cluster') are tolerant and some varieties ('Northern brewer') are symptomless carriers.

In varieties of the "Golding" group, the first symptoms appear early on young leaves on bine and shoot as a veinal mosaic which later becomes a widespread mosaic of yellow spots, which attracts attention, especially on lower leaves. Borders of bine leaves roll downwards to become fist-shaped, the leaves become brittle, the apical nodes are shortened, the vegetative apex loses its ability to climb, falls away the training string and decays. The shoots are short, the cones are small in number or none are formed. The root system gradually decays and the plants die in two to three years.

In the varieties of hop planted in Czechoslovakia, the symptoms of this virosis are considerably different; they appear much later, usually in July, i.e. in the period, when the plants reach the top wires. Diseased plants are identified by their weakness; they have dark bine leaves which tend to roll inwards, becoming fist-shaped. The shoots are short, only a few cones, or none at all, are formed. The leaf colour changes typical of the varieties of the "Golding" group, are rarely seen. Quite exceptionally veinal mosaic appears on lower leaves on the shoots. The symptoms are mainly limited to a moderate rolling of upper bine leaves.

This virosis is transferable by inoculation to indicator varieties 'Petham Golding', 'Early Prolific' and 'KAV'. At the present time, hop mosaic has been successfully transferred only to *Nicotiana clevelandii*.

141

Fig. 59. English mosaic.

Protection depends mainly on the removal of plants with evident symptoms of the disease, within the framework of negative selection. The freedom from the disease of basic breeding material is verified by the use of the above-mentioned indicator varieties.

The virus of hop mosaic was identified serologically in Czechoslovakia as apple mosaic virus being one among the types of Prunus ringsport of the virus (ALBRECHTOVÁ, CHOD, KŘÍŽ, FENCL, 1979).

N e t t l e h e a d . The causal agent of this disease is *Humulus virus 2* (Duffield, Smith). This virosis mainly affects the variety 'Fuggle' planted in Great Britain. Trained shoots grow more slowly than normal, their pale yellow leaves roll upwards becoming boat-shaped. Later symptoms are short shoots growing sharply upwards. The vegetative apex loses the ability to climb and falls away from the training string. The leaves are yellow, brittle and rolled into a boat shape. This disease has not yet appeared in the productive hop stands of Middle Europe.

Bacterioses

Bacterial diseases occur in all hop-growing regions and their injuriousness depends on soil and climatic conditions as well as on agrotechnical procedures.

B a c t e r i a l c a n k e r o f t r a n s p l a n t s (c u t t i n g s) . This is a rare disease, caused by *Agrobacterium tumefaciens* (Smith and Towns) Conn. It mainly attacks the young wood, producing tumours up to a few cm in diameter. The disease can appear on other underground organs of the hop plant, but it has not been found on the above-ground parts.

The prevention of this disease relies on situating the hop garden on dry ground and on soil well-prepared for the planting of the crop. For rooted cutting nurseries, as well as for direct planting, only healthy, undamaged seedling must be used.

B a c t e r i o s i s o f c u t t i n g s a n d s h o o t s. The causal agents of this disease are the bacteria present in all types of soil. Infection starts at the sites of mechanical damage to the underground parts of the plant or in cases of new-wood infected from the internal cavity of the bine. The pith of transplants attacked by bacteriosis first shows brown spots. As the disease develops the neighbouring tissues become brown and decay. The harmful effects of the disease are intensified with secondary attacks of parasitic or semiparasitic fungi (e.g. *Fusarium* spp., *Verticillium* spp.). This secondary attack often obscures the symptoms of the primary disease. Diseased cuttings fail to grow or their grafts die just after budding. Shoots growing from partially infected transplants cease to grow and their apices die.

The hop garden may be protected against bacterioses by the planting of healthy seedlings and therefore any seedlings or rooted cuttings with brown tissues around cuts are systematically excluded. The probability of attack may be limited by cutting the bines close to the ground in the autumn and by the early cleaning of the hop garden.

Mycoses

From the taxonomic point of view the causal agents of mycoses belong to a few groups of fungi which are very different from each other. Some of these organisms can grow and develop only on living tissues of the plant, these are the so-called obligate parasites; others, the facultative parasites, can utilize the tissues of decaying plants.

The greatest losses in terms of quality and yield are caused by *Pseudoperonospora humuli*. Therefore crop protection measures must be taken against this organism every year, and the number of sprays applied will be governed by the requirements of the particular season. Other fungous parasites are injurious to the hop on a lesser scale or within limited areas of hopgardens. In some cases the damage caused by other fungi has not yet been demonstrated in Czechoslovakia in spite of the fact that in neighbouring countries they are a source of serious economic damages.

Downy mildew *Pseudoperonospora humuli* (Miy and Tak.) Wils.

Downy mildew is an extremely serious hop disease and is present in all hop-growing regions in Europe and elsewhere. In Czechoslovakia it was found for the first time in 1925. Between 1926 and 1977 the diseases caused damage in 35 years which included 7 severe attacks. Only in 17 years did the disease not cause serious economic damages (see Table 37). The occurrence and severity of downy mildew depends on the weather pattern particularly on the frequency and amount of prepicitation. It is most likely to be seen in wet years, when it causes the maximum reduction in quality and yield of the crop.

The first symptoms of attack are evident on young shoots in the spring – their leaves catch the attention by their green-yellow colour. The diseased shoots are dwarfed, their leaves are distorted and hang downwards. Because the internodes are shortened, the leaves are crowded and form a shape reminiscent of an ear of wheat and therefore they are sometimes called spicate. These shoots have leaves with a grey violet coating on the undersurface, consisting of summer sporangia, and visible to the naked eye. Spicate shoots appear in the spring after infection has developed from winter sporangia and they are the main source of further spread of the disease. In summer this form of the disease develops only in exceptional cases.

TABLE 37   Injuriousness of downy mildew (Pseudoperonospora humuli)

Year	Injury rating	Year	Injury rating	Year	Injury rating	Year	Injury rating
1926	3	1939	1	1952	4	1965	3
1927	3	1940	1	1953	2	1966	2
1928	4	1941	4	1954	3	1967	3
1929	4	1942	4	1955	1	1968	3
1930	4	1943	3	1956	3	1969	3
1931	4	1944	4	1957	3	1970	3
1932	2	1945	1	1958	3	1971	3
1933	4	1946	2	1959	3	1972	4
1934	4	1947	4	1960	3	1973	3
1935	3	1948	2	1961	3	1974	4
1936	1	1949	1	1962	4	1975	3
1937	4	1950	3	1963	3	1976	4
1938	2	1951	3	1964	4	1977	1

Note: 1 – severe injury, 2 – intermediate injury, 3 – slight injury, 4 – uninjured.

During the growing season *P. humuli* is spread from summer sporangia (zoosporangia) whose spores carry the infection to leaves, inflorescences and cones. The first symptoms, minute green-yellow spots limited by the veins appear on the attacked leaves. In wet and warm weather these spots increase in size and coalesce so that ultimately the whole leaf is attacked. Subsequently, the spots on leaves become brown and dried. The attack starts and its intensity is usually greatest on the lower parts of the plant. In wet weather the disease spreads to the shoot leaves, and will develop severe infection in spicate shoots.

The most serious economic damage caused by *P. humuli* is to inflorescences and cones. The morphology of the cones provides good conditions for infection by the fungus and for its development. When inflorescences are attacked, especially in the early stages of their development, they become brown and in heavy infections they are shed. When cones are attacked during their immature stage, they remain dwarfed and become hard. When the disease attacks developed cones the cuticles become brown, and after heavier attacks the bracteoles change colour. The bracts are arranged in rows one over the other in the cones so that attacked cones often demonstrate infection in stripes. In Czechoslovakia browning of the whole cone is more frequent than striping. Frequently only a proportion of the cone is discoloured giving brown and green variegation. The cones attacked by downy mildew tend not to close up, their scent is of poorer quality and they are of lower value for brewing, because the total content of bitter substances falls by 25 per cent as does the content of tannins. This reduces the various properties on which the quality of hop is evaluated.

The green bine variety is usually regarded as more resistant than the red bine. Czechoslovak varieties are in general more resistant than those of German origin of which the varieties 'Hallerstauer' and 'Tettnang' are most susceptible to attack by *P. humuli*. Likewise the English varieties of the "Golding" group and certain others are disposed to infection. 'Osvald's clones', selections from populations of the variety 'Armat' and 'Siřem' have the same level of resistance as that of the original populations of the variety 'Žatecký'.

P. humuli is propagated by both sexual and asexual reproduction. These alternatives in its life cycle enable this fungus possible to survive non-favourable conditions and, under optimal circumstances to spread very quickly.

Sexual reproduction is always within host tissues. Oospores (winter or permanent spores) originating from the coalescence of two sexually different hyphae can be produced in leaves, cones and other parts of the hop throughout the whole vegetative period. They are round or

Fig. 60. *Pseudoperonospora humuli*
– mature zoosporangia.

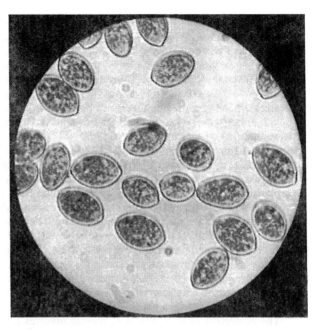

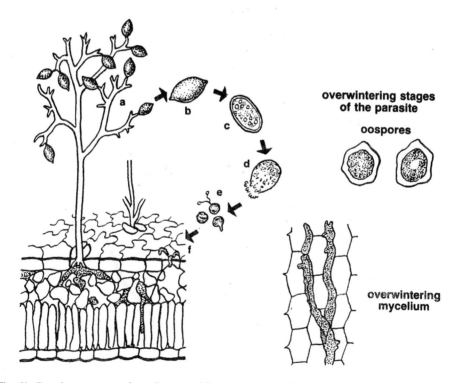

overwintering stages
of the parasite

oospores

overwintering
mycelium

Fig. 61. *Pseudoperonospora humuli*: a – conidiophore with conidia.

oval and measure 36–39 × 34–37 μm. Their surface is covered with a thick membrane which protects them against injury. They are resistant to low temperatures and desiccation. They remain infective for two years. They reach the soil with the remnants of plants and during the winter they are released. In the spring they may infect young shoots growing through the soil.

Asexual reproduction occurs during the growing season. Mycelium grows through the intercellular space of infected leaves and inserts branched haustoria into the cells. Spore-bearing hyphae – sporangiophores grow through the stomata and take the form of antlers (Fig. 61). Summer spores (zoosporangia) are produced on the sporangiophore branches; they are lemon-shaped and measure 17.1 to 36.3 × 13.2 to 26.4 μm. At maturity they are easily separated from sporangiophores and gentle air streams can transport them to relatively distant places. If they reach free water, they release mobile zoospores.

One sporangium usually contains 3 to 7 spores measuring 13 to 14 × 8 to 10 μm of an irregular or oval shape. They are bi-flagellate and the flagella enable zoospores to move actively in water. After a few hours they settle, encyst and start to germinate. The germ tubes grow through the stomata and infect the plant. The zoospores are released at temperatures between 1 and 27 °C but the optimum is between 19 and 25 °C. The length of the incubation period depends on the temperature; the shortest incubation period lasts 3 days at a temperature of 21 to 25 °C. Another factor affecting the production of zoosporangia and their survival is the relative humidity of the air. A dense growth of sporangiophores, carrying the maximum amount of infective zoosporangia develop at a relative humidy of over 90 per cent. Lower humidity encourages fewer sporangiophores and zoosporangia.

The protection of hop against *P. humuli* involves treatment with chemicals and various agrotechnical measures.

The spread of downy mildew can be limited by correct early training of shoots. The hop gardens should be clean and free of weeds which make the environment at lower levels humid. A balanced nutrition, containing all necessary nutriens ensures the healthy growth of the plants and reduces their susceptibility to diseases. It is also important to undertake the early cleaning of hop gardens in the autumn.

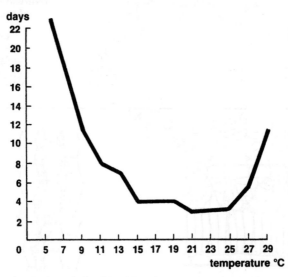

Fig. 62. Temperature-incubation relationship of *Pseudoperonospora humuli* (length of incubation period in days related to temperature).

The chemical protection of hop against *Pseudoperonospora humuli* uses cupreous compounds with oxychloride of copper as an active preventative substance. As another fungicide preparation we can mention Curzate K (oxychloride of copper at 48 % and cymoxanile at 4 %) with a penetrative effect. The preparation Ridomil Plus 48 WP (oxychloride of copper at 40 % and metaxyle at 8 %) is the most frequently applied one among the compounds working through the whole system of the plant.

The fungicides protecting against the *Pseudoperonospora humuli* are applied mostly by means of efficient sprinklers, which reduce the need in labours and the consumption of water and help to better intervention. The water dose ranges from 600 to 2500 l per ha according to the growth of the hop. Leaves have to be treated first of all from under-surface where the infection occurs and this treatment has to cover the whole plant including the top part. Therefore it is important to set the equipment correctly, to choose the width of stroke according to the type of sprinkler and to make the dose of sprinkled material sufficient. This gives the guarantee that the quality of treatment will be good.

The protection of hop against the *Pseudoperonospora humuli* is made during the last years in Czechoslovakia in accordance with its short-time prognosis, which is based on the course of weather and number of diseased plants. The sprinkling days are settled and the need of this treatment is announced. These days are timed so that the plants should be protected in the period of beginning intensive growth, during the period of the formation of shoots, when the plant blossoms, at the beginning of the formation of cones and when the cones ripen. This organization reduces the total number of treatments (formerly 5 to 6, now 2 to 3 interventions).

The application of different fungicides must respect their properties, thus curatively working fungicides are suitable for the period of intensive growth up to the start of blossoming after the training of shoots and the cupreous preparations may be used in the later period finishing with the harvest. The fungicides may be combined with insecticides recommended against the Dowson-hop aphid *(Phorodon humuli)* and red spider mite *(Tetranychus urticae)* in the case of need. It is possible to add also manure for foliar nutrition to liquors sprayed against *Pseudoperonospora humuli*.

The number of treatments can be reduced in those years with a reduced occurrence of downy mildew. Effective protection depends on reliable disease forecasting and timely warning of imminent attack. This is one aim of research in Czechoslovakia and in other countries with a developed hop industry.

Powdery mildew *Sphaerotheca humuli* (DC.) Burr.

Powdery mildew recently became the second most serious disease after downy mildew in Great Britain, the Federal Republic of Germany, Belgium and other countries. In Czechoslovakia, the disease appeared for the first time 1969 in region of Piešťany – Topoľčany in Slovakia and in 1974 it occurred in the region of Žatec.

It is an obligate parasite affecting all above-ground organs of the plant, i.e. the bine, leaves, inflorescences and cones. Young developing organs are especially sensitive to powdery mildew, older organs are naturally resistant. Abroad, the infections starts relatively early, in the second half of May, when small white spots appear on the obverse of leaves. These grow and multiply. Mildew can appear on the undersurface of the leaves though on a smaller scale. Under favourable weather conditions the leaf blades become largely covered with a white floury coating. The attacked tissue quickly becomes dry and leaves may be shed in severe attacks. Diseased buds do not develop so that such plants are partially or totally infertile. Infection is very dangerous when it occurs in cones, which become brown, distort and remain small, their surface remains covered with the white floury coating. Symptoms of powdery mildew attack cannot be mistaken with those of downy mildew but can be similar to those of

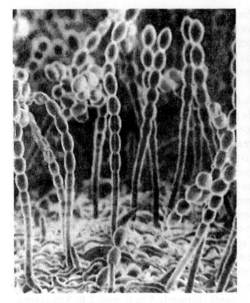

Fig. 63a. A part of sporulating colony of powdery mildew (original dr. D. J. ROYLE, 1978).

Fig. 63b. White mealy cover of powdery mildew on a leaf.

grey rot *(Botrytis cinerea* Pers). Cones attacked by powdery mildew cannot be used for the production of beer, because they have an unpalatable flavour and unfavourably affect the scent and taste of beer. In Middle Europe powdery mildew can attack as late as August when it appears on the youngest leaves and cones.

Plants are treated as soon as the first symptoms of powdery mildew are found and treatment is repeated according to the need. The classical fungicidal sulphur preparations are used in Czechoslovakia, such as Sulikol K (50 per cent colloidal sulphur) as a spray at 1 per cent concentration. When the temperature reaches 25 °C and above, sulphur cannot be used because it could dangerously scorch leaves, inflorescences and cones. This material can be

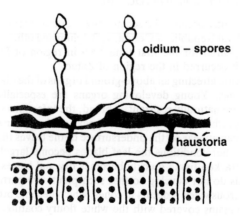

Fig. 64. Powdery mildew.

mixed with organic and cuprice fungicides for protection against attack by *P. humuli*. Of the organic fungicides, Karathane FN – 57 (25 per cent dinocap) is suitable, because it has a curative as well as a preventive effect against powdery mildew.

It is not likely that powdery mildew will develop in Middle Europe to such an extent that it will require regular attention in hop gardens, though it remains as dangerous there as elsewhere.

Fusariosis

The causal organisms are *Fusarium* spp. mainly *Fusarium sambucinum* Fuck, i.e. *Giberella pullicaris*(Fr.) Sacc. It occurs in all European hop growing regions, but in Middle Europe it is found only on swamped heavy soils during wet years. Young wood is particularly sensitive to attack. The hyphae of the fungus grow through and destroy vascular bundles so that the corewood decays. Consequently secondary corewood is developed so that young wood is swollen close to the rootstock. The tissue tends to become brown in colour, the cortical tissues are easily shed to expose small white mounds of sporodochia. The leaves turn yellow, and the plant wilts and successively decays. An infected bine can be easily removed at soil level. The disease can attack one bine whilst the other one remains healthy. The symptoms of the disease are most evident in June and July.

If hop gardens are established in unsuitable places with a high water table or on soils that are not well aerated in wet years, the whole rootstock can be attacked and decay. If the disease is less severe a small number of dwarfed shoots will develop, they are unable to climb, their leaves fail to develop fully, they wilt and shrivel up. If the attack is severe no shoots develop.

Infection can be avoided if land with optimum water conditions is selected for the hop garden. Only healthy planting material should be used; i.e. plants should be not mechanically damaged, and they should be put into well prepared soil. If productive hop gardens are attacked in wet years, it is recommended that the diseased area is ploughed because the cultivation aerates the soil and limits the expansion of disease. Frequent cultivation with a balanced manuring helps to maintain the health of the crop and hinders the spread of the disease. The cleaning of the hop garden in the autumn as well as helping it tidy limits the development of fusariosis.

Verticillium wilt

The causal organisms are *Verticillium alboatrum* R. et. Nerth, and *Verticillium dahliae* KL. Hop wilt occurs in humid hop growing regions in Great Britain and the Federal Republic of Germany but in Middle Europe it is relatively rare. The parasite enters the plant through its roots, wherefrom it grows through the tissues and plugs its vessels. Once the plant is attacked, the first symtoms appear on the lower leaves which become yellow and wilt, subsequently the disease proceeds upwards, the whole plant wilts and the leaves become dried and fall off. Protection is through indirect methods – the hop gardens should be established from healthy seedlings, the correct agrotechnical measured should be used, and the nutrition of the plants should be adequate.

Grey rot *(Botrytis cinerea* Pers.)

This rot attacks plant parts on the ground or weakened ones. Leaves and cones turn brown and a red-brown to grey-black coating develops on their surface. They may be protected by spraying with 0.2 per cent Euparen (50 per cent dichlofluanid) twice after the start of blossoming of 10-day interval.

Pests

Hop gardens had in the past a relatively rich entomofauna. There were many species which depended specifically on hop as their food source as well as many which were polyphagous. But the intensive chemical protection of hop, with the repeated and continuous application of organophosporus insecticides with their broad spectrum of activity has substantially reduced fauna of hop gardens. Many species sensitive to the insecticides or not able to find suitable conditions in the gardens cultivated by modern methods do not now occur at all. Some pests especially aphids and spider mites have developed biotypes with high level of resistance through this continuous selective pressure. The increased resistance of pests to insecticides nowadays represents a serious problem in the protection of the hop.

Lovage weevil (*Otiorrhynchus ligustici* L.)

This weevil was a long time only an occasional pest of hop, using it as a food source during its spring migration from its hatching place. Damage caused by this insect was found for the first time in the region of Žatec in 1964 on the underground parts of the plants. An investigation of its development cycle confirmed that some populations of lovage weevil are adapted exclusively to the hop so that their entire development is confined to the hop garden. Thus it became the primary pest of hop and it is now necessary to be always prepared for its occurrence. It prefers a sandy-loam soil where it occurs most frequenty.

Wintering beetles of the lovage weevil emerge individually from the soil from the end of March onwards. Large numbers emerge when the soil temperature reaches 13–15 °C. These are not nocturnal beetles as weevils usually are, so that they can be caught during the daytime, grazing on young shoots, where they feed especially at the vegetative apices. In cold weather they dissapear into the soil and devour the apices of budding shoots. If they are numerous, they eat most of the shoots, but if they are fewer, their presence is usually not discovered before the vegetative apices of shoots, which have developed after the cut, are found to be damaged. These shoots do not produce many lateral branches. The stands attacked by weevils are unbalanced, their growth and development is considerably retarded and the harvest is consequently reduced.

Less evident, but very serious, damage is inflicted by larvae on the underground organs of hop plants. Larvae are usually found at a depth of 20–30 cm, on the surface of roots and young wood where they eat out channels and grooves in the cortical tissue. They are usually found here at different stages of development. Even a few larvae can cause such severe damage to the underground organs of the hop that after years the plants decay one after another. The stands thus become uneven and, the proportion of gaps in the crop becomes high.

The beetle is approximately 1.5 cm long, broad with a short flat proboscis, slender antennae and broad oval elytra. It has black shiny legs, thick femina, and the end tibiae are very large. Because the elytra are fused and the membraneous wings tiny, the weevil does not fly. The basic black colour with yellow hairs can change according to the colour of the soil. The population consists mainly of females. Males are rare, therefore propagation occurs from non-fertilized eggs. These are usually laid at the end of April. They are almost spherical, of 0.8 mm diameter; at first they are light yellow but later become dark. The females lay eggs in groups at or just below the surface, immediately around the hop plants. The laying period lasts up to 2 months and one female lays 200–600 eggs. The larvae emerge after 15–20 days. They are legless, white, gently deflected and anally contracted. They have a small brown head with well developed mandibles. During their development they cast their skin 7 times and before pupation they are 13 mm long. Pupae have a shape similar to that of the adolescent beetle. They are white yellow covered with short blunt setae. They are found immediately around the hop plant in small oval chambers at a depth of 30–60 cm. Their development lasts 20–30 days.

150

The complete development cycle of the lovage weevil in Middle European hop-growing regions usually lasts 2 years. The larvae overwinter in the first year, and the beetles in the second, because they do not emerge immediately after their transformation but remain in the underground chamber until the following spring. Three-year development has been observed only exceptionally, in certain individuals which originated from eggs laid at the end of the laying period. In such a case the larva overwinters for two years and the beetle for one year. The lovage weevil is broadly polyphagous but in the hop-growing regions appears to prefer hop. The amount of precipitation in the region of Žatec and moderate winters make conditions there favourable for its occurrence. The population density of this beetle is controlled by heavy spring rains because many beetles are drowned in puddles. Among the bioregulators are the parasitic fungi *Beaveria globulifera* Speg. and *Isaria farinosa* Diks., as well as certain animal parasites such as the small wasp *Tomicobia (Ipocoelius) rotundiventris*. Among the important predators are the beetles. *Calosoma inquisitor* L. and further some carabids *(Carabidae)* as well as gadflies *(Staphylimidae)*.

Protection concentrates on the beetles. In the period 1968–1976 chemical protection based on γ – HCH (as lindane WP 80) applied by sprinkling or spraying during the period when beetles emerged. The rate was 5 kg per ha. Because its efficiency decreased and many side effects developed such as the production of tumours in young wood, alternative insecticides came into usage. Beetles can be exterminated by the carbamate insecticide Furadan 75 WP (75 per cent of carbofuran) applied by sprinkler or by strip spraying at the rate of 1 kg per ha. It has no known side effects, is rapid in action and its effects are long-lasting. Because the development of the lovage weevil may take up to three years it is necessary to repeat the treatment at least every three years.

Flea beetle (*Psylliodes attenuata* Koch)

This is the most common and most widespread pest of hop. Its injurious activity is twofold: in the spring it causes damage to the budding shoots and leaves of young plants, and in the summer to cones and branch leaves. The beetles create large holes in the vegetative apices and leaves of young shoots. Damaged leaves are perforated and may even become skeletal, when only the veins are left. Seriously damaged shoots do not grow, remain short and decay. The summer generation of the beetle damages the young leaves of lateral branches in July and August but does most damage to cones. The beetles eat the cones as well as the spiral. Damaged cones do not grow. The summer generation causes damage to plants growing around the poles and it also attacks fallen bushes whose cones are totally destroyed within a few days. Flea beetle was important pest of the hop in the past but recently in consequence of an intensive chemical protection it has become of only secondary importance.

The beetle is 2–2,5 mm long, oval, green-black, with a metallic sheen. It has long slim antennae with the first five segments rusty red and the last five black. Its legs are of the same colour as its antennae. The femina of the hindmost pair of legs are strongly developed and modified for jumping. Flea beetle eggs are small (0.43 × 0.25 mm), oval, light yellow, with small polygonal markings on their surface. Larvae are threadlike, approximately 4 mm long, whitish with a yellow head, scutellum and pygidium and with three pairs of short legs. The pupa is egg-shaped, 3 mm long, 1.5 mm wide, and very contracted at the ends. It is white until just before the emergence of the beetle when it becomes brown. The sheaths of wings, the elytra, legs and antennae are evident on the pupa.

The beetles emerge from their winter retreats when the days become warm during April and the beginning of May. They are thermophilic and heliophilic, therefore they remain hidden in the soil when the weather is cold and clouded. The feeding period lasts for 1–2 months, after which the beetles mate and the females start to lay eggs around the end of May. One female lays approximately 150 eggs, placing them in the soil around the plant roughly at a depth of

5 cm. Approximately 10 days later the larvae emerge. At first their feeding forms tunnels in the fine roots, then they reside in the soil and consume only the finest roots but do not cause serious damage. During their development lasting 25–49 days they cast their skin twice then move deeper into the soil where they pupate in a small chamber. After 3 weeks the beetle emerges from the pupa. Thus the flea beetle has only one generation per year. The adult beetles live for a few days on the youngest leaves and at the end of October they prepare for winter. They find their winter shelter mainly at about 10 cm deep in the soil of the hop garden, or under plant remains.

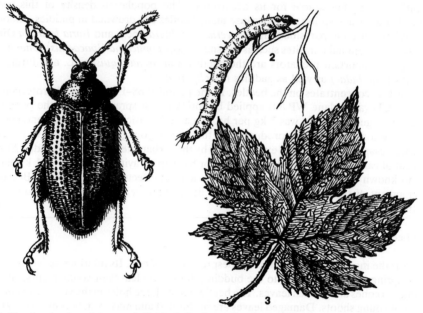

Fig. 65. Flea beetle: 1 – adolescent individual. 2 – larva, 3 – damaged leaf.

The population density of this pest is controlled to a large extent by abiotic factors, and in particular alternating thaws and frosts are critical for overwintering beetles. Eggs and larvae are sensitive to drying and if the soil moisture falls below 20 per cent, the eggs die. Among the enemies of flea beetle, a species of *Perilitus bicolor* is known to be a parasite.

Protection against flea beetle is not usually necessary, because the regular application of insecticides to control the Damson-hop aphid over a long period reduced the population density of this beetle which had previously been a very serious pest. In exceptional cases methidathion (as Ultracid 40 WP) can be successfully applied at a rate of 2–3 kg per ha or carbofuran (as Furadan 75 WP) at 0.5–1.0 kg per ha.

Damson-hop aphid (*Phorodon humuli* Schrank.)

This is the only aphid living on the hop but it is an economically important pest in all hop-growing areas in temperate and subtropical zones. The hop is the secondary host plant of this aphid; alate females arrive on the crop from the second half of May. They settle on the youngest apical leaflets and without mating they bear living larvae which develop into wingless females. Under favourable conditions up to 8 generations of aphids are produced to damage the hop plant during its growing season. Direct damage is due to the removal of plant nutrients by the sucking aphids and due to the development of saprophytic fungi of the genera *Capnodium, Cladosporium* and *Alternaria* which produce a dark coating as they grow on the sweet faeces of aphids. This coating hinders the assimilation and respiration of the leaves. When there is a small number of individuals they are usually on the underside of the leaf where the veins branch. Such colonies grow and finally cover the whole underside of the leaf. Loss of plant sap sucked out by many individuals is considerable and represents a serious interference in the metabolism of attacked plants. The aphid, as a non-migrant species, takes up nutrients continually.

Damaged leaves, seen from below first become transparent and later, especially if the number of attacking aphids is large, the edges roll inwards forming a fist shape. The growth of the plant becomes retarded and eventually stops. The individual bright spots of honeydew which appear on the obverse of leaves, soon spread to produce a compact adhesive coating. Heavily attacked plants cease to grow, do not produce side branches and develop a peaked shape. Later, if the aphids are exterminated, the damaged areas remain empty without side branches, shoots and cones and this considerably lowers the yield. If aphids penetrate into the cones then the whole yield is always destroyed. The empty aphid skins and dead bodies together with their excrement accumulate in the cones which then become blackened by fungal growths.

As shown in Table 38, aphid development starts on the primary host plants – plums and blackthorns, where the long oval shiny black eggs (0.6–0.8 × 0.3–0.4 mm) overwinter. They are usually laid individually on shoots, in leaf axils or on rough places of the cortex. Dark green larvae emerge from the eggs before bud burst at the end of March. These become generative females after four skin castings.

TABLE 38   Life cycle of Damson-hop aphid

	Month											
	I	II	III	IV	V	VI	VII	VIII	IX	X	XI	XII
Winter eggs	———		———									
Fundatrices			———		———							
Fundatrigeniae					———		———					
Migrantes alatae						———		———				
Virginogeniae						———			———			
Gynoparae										———		
Males										———		
Oviparae										———		———
Winter eggs										———		———

The shape of the body of this female is considerably different from that of all other females appearing on the primary host or on the hop. It has no portuberance on the front and first antenna element which is typical of all other generations. Its body is also much larger, 2.0–2.3 mm long, 1.3 mm wide, oval, annally dilated; its pygidium is very well-developed. It is light green with a narrow longitudinal dark stripe on its back. Its antennae have five elements and are up to one third of the length of the body. It has produced 125 larvae in glasshouse experiments and 81 larvae according to the observation in nature. Wingless females, known as fundatrigenies emerge from the larvae produced by this female. Usually only complete fundatrigenous generation develops on the stone-fruits. The fundatrigenies are longitudinally oval 2.3–2.7 mm long, paler yellow-green, than the founder-female,. with three darker longitudinal stripes on the back. Its frontal protuberances, so typical of the hop aphid, are fully developed. Its antennae have 6 elements, the stylet (proboscis) reaches to the second pair of legs. Its fertility, according to observations in nature ranges from 56 to 65 larvae. Its descendants usually become winged (migrantes alatae). Only a small proportion of larvae, from the youngest shoots of plums, develop into another very weak generation of fundatrigenies. Thus, the protracted development of fundatrigenies on plums also gives rise to alate females in the second and later generations, but these are very small in number. The migration of alate females to the hop is similarly protracted.

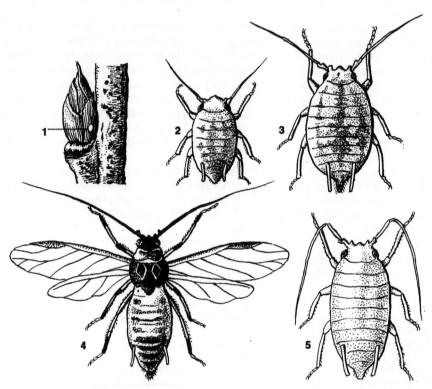

Fig. 66. Damson-hop aphid: 1 – plum bud with an aphid egg, 2 – larva immediately after emergence from the egg, 3 – viviparous female, 4 – alate female, 5 – viviparous female from the hop.

The winged females are 1.5 to 2.2 mm long, light green, later grey, with dark 1.8 mm antennae of 6 segments. They have a dark stripe on the prothorax and three dark green spots on the thorax. The frontal protuberances are slender, parallel and clearly developed. Its pygidium has transverse dark stripes and 3–4 dark side spots. Its wings are long and broad and at rest they take the shape of a roof. The migration of alate females to the hop, under Middle European conditions, starts during the second half of May, culminates in the first third of June and ends during the first week of July. The last alate aphids may; however, be found on the plants even at the end of July. Migration occurs at a minimum temperature 17 °C at the leaf surface and the wind speed must be less than 1 m. p. s. The wind carries aphids over long distances but for migration from plant to plant they use active flight. Finally they settle in the folded upper leaves.

The hop plant is only the secondary host plant of this aphid, where the alate females give birth to approximately 21 larvae which will develop into wingless virginogenous females. The number of generations on the plant during its growing season ranges between 5 and 8. The virginogenies, wingless parthenogenic females, are 1.6 to 2.0 mm long, are pale yellow-green in colour, and have no green stripe along their backs. They have the frontal protuberances as well as the protuberances on the first segment of the antenna very well developed. Under Middle European conditions the development of virginogenies takes eight to twelve days. Their fertility depends on the temperature as it does in the other forms, and ranges in nature at about an average of 86 larvae.

During the cone harvest, at the end of August, alate females (gynoparae) develop on the hop and migrate back to plums. Their body shape is not very different from the body of migrates alatae. On the plum they produce larvae which become oviparous females. At the same time the wingless females remaining on the hop produce larvae which become alate males. These also migrate to plums, where they copulate with the oviparous females, which then lay the winter eggs. The alate males are smaller than the gynoparae, they are 1.35 to 1.6 mm long. Their wings are longer, their antennae somewhat more elongated. The oviparous females on plum are different in body shape from the parthenogenic females. They are longer with a much better developed pygidium. The basic yellowish colour is mainly covered with dark irregular spots. The frontal and antennal protuberances are well developed and are lighter in colour than the head. The oviparous female does not lay more than six eggs. The first eggs can be seen at the end of September. If the weather conditions are favourable, eggs can continue to be laid to the end November. The eggs are mainly found on plums and blackthorns; they are very tough and, under Middle European conditions more than 90 per cent survive the winter.

If the mean fertility of the Damson-hop aphid, in its various forms, is considered, the total number of generations can be determined and from it the biotic potential. If the lower values obtained by observation are used, then the following results are obtained.

This indicates that one egg overwintered on the primary host can theorically produce 4536 alate migrantes (winged females) and after they have migrated to the hop there could be at least seven further generations. Under these conditions, the biotic potential of the hop aphid could reach 21.86^6. This enormous power of multiplication is balanced by a corresponding resistance of the environment, so that over-population will not occur. The extermitation coefficient (g) calculated according to BREMER'S method which indicates the mortality required to maintain a fixed level of population is 98.8 per cent for the fundatrigenous generation and 98.2 per cent for the virginogenous generation. This tremendous biotic potential of the hop aphid implies that the number of alate migrantes arriving on a hop plant need not necessarily be very high in order to endanger the plant. On the other hand the high value of the extermination coefficients shows clearly that any reduction in crop protection procedures would soon induce a new danger to the plant due to the overpopulation of aphids. Abiotic factors, especially temperature, affect the build-up of the hop aphid directly influencing the speed of its population dynamics as well as indirectly influencing, at the same time, the population dynamics of its enemies. The most important predators of hop aphids include larvae of ladybird beetles (Coccinelidae), larvae of drone flies (Syrphidae), and the larvae of certain neuropterous (Neuroptera) and rapacious bugs (Heteroptera). The complicated relations between the hop aphid and the biotic and abiotic factors of the resistance of the environment need further analysis. In the past the predators kept the hop aphid under control. BLATTNÝ explained the periodicity in the calamitous occurrence of hop aphids which was typical of the period from 1900 to 1925 was caused by the intervention of natural enemies, because at that time the chemical protection of hop was not widely used. More recently the main controlling factor has been chemical protection and the involvement of predators is no longer decisive. The long term application of insecticides, with a broad spectrum of efficiency, required for economic reasons, also affected the favourable entomofauna of the hop gardens and lowered the biotic resistance of the environment in the hop growing regions. In this situation, characterized by the non-existence of natural enemies, the high biotic potential of the hop aphid is increasingly relevant and the aphid itself consequently becomes a very dangerous hop pest.

Protective measures against the aphid should be used every year, not only on the hop but also on the primary hosts. The aphid is historically the first hop pest against which, as early as the 1920s first protective measures were taken. Until the 1950s contact insecticides of vegetable origin (nicotine, pyrethum) were mainly used. These were gradually replaced with more efficient materials based on organic compounds of phosphorus, the advantage of which lies in their systemic effect. The results obtained with systemic organophosphates were so good that it seemed that the chemical fight with the aphid was solved forever. But after ten years of the successful application of sprayed and sprinkled organophosphates a calamity with aphids appeared in the hop.planting region of Žatec (Czechoslovakia) in 1967, when the insecticides based on thiometon failed. It was soon confirmed that the hop aphid had developed resistance to all insecticides successfully applied up to that time. Detailed studies of this resistance over the succeeding years showed that it is a very complicated problem because aphids become resistant to preparation of quite different chemical structure which had not yet been used at all in practice (HRDÝ, KULDOVÁ, 1970, KŘÍŽ, 1973, 1975). The development of resistance to insecticides makes chemical protection very problematic. The short useful life of an insecticide must be considered and continuous attention should be paid to the selection and testing of alternative compounds.

Only those insecticides are suitable for use on hop which can overcome the spectrum of resistance present in the contemporary generations of hop aphids. A list of suitable insecticides and their application is included in the Guide to Plant Protection published by the Czechoslovak Ministry of Agriculture and this guideline is closely followed. The preparations and their application rates recommended for use against hop aphid during the period 1976–1980 are shown in Table 39. In hop gardens with broad spacing sprayed insecticides are usually used and applied with efficient sprayers. The sprinkled insecticide Terra Sytam (50 % dimefox) is usually applied only to hop gardens with narrow spacing which are non-accessible to modern spraying equipment.

TABLE 39   Insecticides recommended for use against Damson – hop aphid

Compound	Active substance	Application rate per ha	Harvest interval (days)
Cybolt	10 % of flucythrinate	0.3 – 0.4 l	28
Vaztak	10 % of alfa-cypermethrin	0.45 – 0.6 l	21
Karate	5 % of cyhalothrin	0.4 – 0.6 l	14
Decis	2.5 % of deltamethrin	1.0 – 1.2 l	14
Talstar	10 % of biphenate	0.6 – 0.75 l	21

The effectiveness of protection depends on the use of a suitable preparation but very important also is time of application. Because the insecticide used against aphids is usually applied with the fungicides against *Pseudoperonospora humuli* the first treatment is given relatively early, at the beginning of June, when 50–100 larvae of wingless females can be found on the mean hop leaf. Up to the 15th July, when the migration of winged aphids is almost over it is necessary to apply three or four sprays. The extermination of aphids should be finished by the second half of July, because this is the time when cones are formed. It is also very difficult to exterminate them later and in any case, if the cones are attacked and the aphids are then exterminated, their remnants in the cones seriously lower the quality of the hops.

Correctly prepared sprayers and the proper application of the treatment, especially with

Fig. 67. Application of pesticides in the hop garden by means of a powered sprayer.

regard to spray pattern and recommended speed are most important to achieve maximum effectiveness of all protective interventions. Without this the aphids will survive in the apical parts of the hop plant. All insecticides used for the extermination of aphids are toxic to humans and therefore it is essential to observe the respective safety regulations. It is also necessary to observe the harvest interval which is the shortest allowable time between the last application and harvest.

Red spider mite (*Tetranychus urticae* Koch)

This is a very dangerous pest wherever hops are grown. After 1956, when organophosphates began to be used against red spider mite in Czechoslovakia, its importance decreased. However during the dry and warm years (1975 and 1976) local populations, highly resistant to organophosphate insecticides, were found in the hops in the region of Žatec (KŘÍŽ, GESNER, 1976). Red spider mite is typically polyphagous causing damage to many other plants, including beans, cucumbers, strawberries, cotton, fruit trees and various decorative plants.

The first clear symptoms of damage to hop plants caused by red spider mite are usually found in June, but after a dry and warm spring damage can be found earlier. Sometimes severe damage to shoots caused by overwintered females has been found. The shoots of such plants remain dwarfed and the leaves develop only a little and are dry, so that the final impression is of the *Peronospora* infected spike shoots. But this level of early damage is exceptional. Usually red spider mite damage appears in June, on low leaves on the bines as tiny individual spots of damaged tissue passing smoothly into the normal green. The upper surface of these spots is slightly convex, giving rise to the name "blisters of red spider mite". In dry warm weather the red spider mite spots rapidly grow in number. Adult females migrate to higher leaves where they cause new spots. The spots on lower leaves join together and the leaves first become yellowed and later paper-grey in colour. Heavily damaged leaves become completely dried and are shed. Similar damage is caused by red spider mite to the leaves on shoots. If protective action is not taken in time, the red spider mite moves to the cones. Early-attacked cones do not develop, they remain small, turn red-brown, dry out and are removed by the wind. Later attacked immature cones turn brick red and their value for brewing is very low.

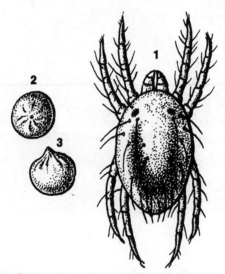

Fig. 68. Red spider mite (*Tetranychus urticae*) in adult individual (1) and its eggs (2.3).

Large numbers of eggs can be found on leaves damaged by red spider mite as well as colonies of all other stages of its life cycle. An egg of red spider mite is spherical and is approximately 0.13 mm in diameter. Newly laid eggs are transparent but as the embryo develops, they turn yellow and later become yellow-orange. A recently hatched larva is also almost spherical and colourless, with 6 legs, bright carmine red eyes, and it measures 0.15 × 0.11 mm. Once it begins to feed it turns grey-green in colour. After a short sucking period, the larvae enter a phase of temporary rest, after which they cast their skin and pass into the second larval stage, the so-called protonympha, which has four pairs of legs. This, in turn enters a rest period and, after casting its skin, the third stage appears, the so-called deuteronympha. After a final rest period the adult individuals enter the adult stage. The female is approximately 0.48 mm long and 0.28 mm wide. Its oval body is not clearly articulated. It has long colourless hairs on its convex dorsum set in six transversal rows. It has a piercing and sucking apparatus. The final article of the mandible is modified into a long piercing seta. The eyes are red, placed on sides of the cephalothorax which is separated from the rest of body only by an indistinct seam. An adult red spider mite has 4 pairs of single articulated legs. Last article is a sole (track) with a comblike formation called the empodium, the shape of which is important for the identification of the species. The male appears smaller and slimmer, 0.35 mm long and 0.18 mm wide, but has the same morphology as the female. Both, males and females have spinerets which produce the cobweb around their colonies intended to protect them against harmful external factors. The body colour changes according to the season. After the final skin casting the adult individuals are colourless. But as they start to take up nutrients, the content of the digestive organs shows as a green colour through the body. The colour of the body at that time is light yellow with small dark spots along the sides. These spots are of irregular shape and size. In the autumn, in consequence of food intake, individuals become orange to brick-red in colour, the typical colour of winter females. As soon as the overwintered females begin to suck green plants in the spring, the red colour fades and the body becomes orange-green.

In fact, only adult females, fertilized in the autumn, survive the winter. They pass the winter under plant remnants in the hop garden, in dry grass stems, and sometimes in fissures in the soil. They are always found grouped in large colonies. Though these females are resistant to low temperatures, they are less likely to survive wet and warm winters. The females usually leave their winter retreat towards the end of March, as soon as the air temperature reaches 10–12 °C. At first they live on different weeds [(e.g. chick weed (*Stellaria media*), dead nettle (*Lamium* sp.), speedwell (*Veronica* sp.)], but as soon as the hop shoots emerge, they move on to them. In glasshouse experiments the average fertility was higher than 150 eggs from each overwintered female, but some of these are not fecundated. The fecundated eggs produce females, the non fecundated eggs males. In the next generation, the eggs from fertilized females produce both males and females, the eggs from non-fertilized females however only males. The ratio of the sexes in the population changes according to the season. In the spring females are prevalent, but later the ratio becomes balanced. The length of the embryonal and postembryonal development is mainly governed by the temperature. Therefore, the temperature also governs the number of generations living on the hop over the season and consequently it controls the amount of red spider mite damage. A lower relative atmospheric humidity also has a favourable effect on the number of eggs laid and on the length of life of each individual.

Under Middle European conditions, there will usually be 5 to 10 generations of red spider mite per year (BLATTNÝ, OSVALD 1950). The rapid development and high fertility of individuals are the basis of its high biotic potential which is fully demonstrated only in extremely dry and warm years. The red spider mite spreads very quickly from severely infested centres in which passive dissemination by the wind plays an important role. This lifts the whole, spun, cobweb from the plant enclosing all of the postembryonal stages. Among the enemies of red spider mite are many minute bugs, rover beetles (staphylinids) and thrips, but an important predator is a small (approximately 1 mm long) ladybird (*Scymnus punctillum* Weise) and its larvae.

Protection against red spider mite depends on chemical and agrotechnical interventions. It is necessary to ensure that the hop garden and its immediate neighbourhood is free of those weeds on which the pest could live during early spring or after the harvest and the remnants of which could serve as a winter retreat. Chemical protection has to be applied immediately when the first blisters are found on the lower leaves.

Organophosphate insecticides have been applied in the protection of hop for a long time so that resistant populations of red spider mite now exist. Therefore, at present, only acaricide preparations are useful. The following acaricides are most effective. Mitac (20 per cent amitraz) and Milbol EC (25 per cent of dicofol), both at 0.2 per cent and Plictran 25 W (25 per cent cyhexatin) at 0.1 per cent. Because these, and similar compounds are contact acaricides, the method of application is vitally important, the leaves must be well covered by the spray and the rate should be not less than 2000–3000 l per ha. The effectiveness of the whole treatment must be assessed and if required, further protective measure should be taken. Good results are likely if chemical protection, is applied in good time, during the initial (i.e. weak) occurrence of the pest. Later, if the beetle has got up to central and higher leaves, the spraying operations are difficult and costly. Chemical treatments should be repeated several times. Once the red spider mite has penetrated the flowers or cones, it is no longer possible to prevent the development of damage. Therefore chemical treatment should be successfully completed by the first half of July, before the hop starts to blossom.

Noctuid moth or rosy rustic moth (*Hydroecia micacea* Esp.)

Caterpillars of this moth are only occasional pests of the hop. Under normal conditions they are found in root-heads of various hydrophytic plants such as water-reed, horse-tail, coltsfoot, clay-weed and dock as well as in the rootstock of the wild hop. In certain special conditions this moth can become an important pest of the hop.

The first symptoms of damage by the caterpillars usually appear at the end of April, at the time of the emergence of the first shoots. Unusual shoots appear in which the vegetative apex is wilted and hangs down or is totally dried out. The shoots can be easily lifted from the soil and in their base will be found a tunnel filled with detritus and excrement. If the shoots are newly attacked and not yet dried out then a minute (0.5–1.0 cm) red-violet caterpillar may be found inside the tunnel. An entry hole will be found in the lower part of the shoot on the level of the soil surface. The caterpillar lives inside the shoot for only 4 to 8 days, leaving it as soon as the upper part wilts and attacking another. So each caterpillar can destroy many shoots. Before the hop shoots are trained the attack can be easily overlooked, but is often found much later when the vegetative apices of trained shoots wilt. It seems that the caterpillars are less damaging after the extra ploughing; adolescent caterpillars do not then attack more bines but migrate to the underground parts of the plant, to rootstocks and roots, where they bore a maze of corridors filled with detritus and brown excrement. When there is a large number of caterpillars then the underground parts of the plants may be totally destroyed. This damage will become evident in the new year at the time of cutting. The adult caterpillars leave the roots at the beginning of June and pupate in the soil in the immediate neighbourhood of hop plants. A moth emerges from the pupa at the end of July or during August, but this is rarely found in nature.

Protection involves the rise of agrotechnical and chemical measures. The basis of this is the maintenance of hop gardens in good agrotechnical condition, without weeds, especially in the borders and within the rows. The most important requirement for the eradication of the noctuid moth is thorough weeding, especially the extermination pf couch grass.

Fig. 69. Application of pesticides to the hop garden by aircraft.

It is possible to get rid of the caterpillars by removing the fading shoots during the training of the hop, but the removed shoots must be immediately taken away from the hop garden. In an attacked hop garden, it is useful to leave at least 2 spare shoots which can then be removed immediately before the extra ploughing. From July to harvest, i.e. in the period of egg laying it is necessary to keep the stands weed-free. For this it is possible to use soil herbicides based on atrazine, which should be applied after extra ploughing. The acaricide of choice applied as a spray or as a drench, is Furadan 75 WP (75 per cent carbofuran) at 1 kg per ha. The treatment may be applied after cutting as the shoots are emerging. When the number of caterpillars is very high, Furadan 75 WP is applied at a concentration of 0.06 per cent and at the rate of 0.5 l of the liquid to each plant. Such chemical protection should be used to only a very limited extent.

The c o d l i n g m o t h (*Pyrausta nubilalis* Hüb.). The caterpillars of this pest of hop gardens overwinter in remnants of bines not taken away. Therefore machine-harvesting and the early removal of crop remnants after harvest limits the numbers of caterpillars which create tunnels in the cores of wood and bines.

W i r e w o r m s (*Agriotes* spp.). These larvae of elaterid beetles cause much damage in plantations and in new hop gardens where they make deep holes in transplants and young wood and sever young budding shoots. They may be exterminated by the following preparations: Mocap 10 G (10 per cent ethoprophos), Dyfonate 10 G (10 per cent of fonofos) and Furadan 10 G (10 per cent of carbofuran) used at 20 kg of granulate per ha in an existing hop garden or 6 to 7 kg of granulate per ha placed in the planting holes at planting.

H o p s t r i g m i d g e (*Contarinia humuli* Tolg.). Larve of this small gnat occasionally cause damage to the hop by feeding on the strigs of cones. The insecticides used nowadays against aphids limit the occurrence of these larvae.

G h o s t s w i f t m o t h (*Hepialus humuli* L.). The yellowish black-spotted caterpillar with a black head causes occasional damage to the underground organs in June and later. Attacked plants wilt and their bines die. Protection is not deemed to be needed in view of its very limited occurrence.

Protection methods applied to the hop

Chemical protection of the hop is not the only available, despite the fact that it is nowadays the most common. *Integrated protection* involving all methods economically, ecologically and

161

toxicologically suitable for keeping pests and diseases below the threshold of damage, become gradually more and more important as it uses a conscious application of natural limiting factors. This method primarily involves the usage of those factors which promote the natural mortality of the pest. Pesticides in this environment should, therefore, be used only in cases of unavoidable need. The indirect methods of protection are mainly preventive. They concern agrotechnical principles, i.e. the correct ploughing of the soil, the provision of the optimum nutrition to the crop and, finally, the selection of varieties resistant to the various diseases (e.g. *Verticillium* wilt, and downy and powdery mildew). Breeding for resistance against *Pseudoperonospora humuli* is successful in several countries and is pursued in Czechoslovakia. There are also phytoquarantine regulations covering different legal and practical measures aimed at stopping the spread of dangerous pests and diseases to uninvaded territories. The direct methods of protection concern the mechanical, biological and chemical methods of the fight against harmful factors.

BREEDING AND PROPAGATION OF THE HOP, THE PRODUCTION OF TRANSPLANTS

The breeding of new varieties

Two types of input material are used in the breeding of new hop varieties.

The first type includes *plants selected from stands of regional varieties*, the biological and economic properties of which had developed under natural conditions and also the negative and positive selections of progressive hop growers in their own hop gardens. Despite these selective interventions the regional varieties, developed in a broad population consist of biologically and economically different types and therefore they represent good material for the breeding of new clonal or population varieties. Both of these selections are from the original regional variety and so they do not differ substantially, especially with regard to the quality of their cones, from the original regional varieties. This selection method was used for the breeding of all commercial varieties accepted for planting in Czechoslovakia. These varieties are identical with regard to the quality of cones but differ in other economic properties as was mentioned in their description under the chapter concerning the biology of hop.

The second type of breeding material results from *the crossing of individuals with different properties*. The resultant hybrids usually demonstrate the heterosis effect. Cross breeding involves the use of individuals with significantly modified genetic properties, which in some cases have arisen, by mutation, e.g. by chemomutation, radiomutation, or have been created by polyploidy. Hence, it is possible to produce an individual with entirely new properties, for example with higher resistance to disease. Such plants are used for cross-breeding in order to produce hybrids with the required properties. The hybrids are grown on in breeding nurseries where they undergo rigorous positive selection.

The best individuals are clonally propagated and are submitted for examination to the State testing station in Žatec. Here their main biological and economic properties are experimentally examined and compared with those of the existing certified varieties. The best of the new clones will finally be included in the Czechoslovak list of permitted varieties.

Maintenance of stocks by breeding

Stocks of the permitted Czechoslovak varieties are maintained further by breeding. Such breeding is necessary because of a gradual deterioration of the varieties in cultural practices. This deterioration is twofold: genetic and non-genetic. One particularly important sudden genetic change is brought about by mutation. Among the non-genetic changes is that caused by unfavourable external conditions during propagation and cultivation. Any undesirable deviation in the stand increases with the repeated propagation of that variety.

The great advantage of vegetative propagation is that the genetic basis of the maternal plant is transferred, without change, to the new plants. However it has a serious imperfection in that the vitality of the variety tends to decrease. This reduced vigour is evident in the metabolism, growth and productivity of the variety. Likewise, the resistance of each plant can decrease and the number of plants attacked by virulent diseases will increase. The reduction in vigour and disease resistance will not be the same for all plants, in the stand so that the yield and health of many of them will be unaffected. Therefore the positive selection of best individuals in the stand, together with a strict removal of unwanted individuals, is the basic method of breeding for maintenance. Using this method 10 to 30 per cent of plants are selected from the total in the stand of the bred variety. This select group is further improved by repeated annual negative selections, i.e. the removal of unsatisfactory plants. The basic material thus produced is used for further propagation.

Propagation of hop

Material suitable for the vegetative propagation of hop is obtained from the plants produced by breeding for maintenance or from authentic stands of maternal varieties in hop-gardens. Negative selection is used in the propagating stands whereby weak and diseased plants as well as those of different varieties and types are rogued out before the stand is authenticated. Material to be used for seed production can be used only after the authorization procedure which is repeated in the maternal hop-garden containing the propagating stand every year. The stands which fail the authorization are excluded from the scheme for the propagation of planting stock. When larger surface of maternal hop-garden has got an authorization, than foreseen by the plan, then the production prefers new varieties and higher degree of propagation. The Czechoslovak State Standard CSN No. 46 3790 – Hop Planting gives details of the authentications procedure for propagating stands. This procedure has been used in Czechoslovakia for a long time. Its basis was fixed in 1956 when the first Czechoslovak State Standard for hop planting was issued. The city agency of the central agricultural committee in Žatec is charged with issue of authorizations.

The stands of bred and local varieties of hop included in the list of permitted Czechoslovak varieties with their propagation grades undergo the authorization procedure.

Authorization cannot be sought for those stands, where different varieties, clones or propagation material are grown under the same hop-yard construction, or stands heavily affected by harmful factors, or stands with refused authorization in previous years because of the presence of viral diseases or contamination with another varieties. Only one stand can be included in one application form (one stand under one construction). The stand must be properly protected against diseases and pests. The applicant must do a negative selection in the propagation stands, before the first inspection of the stand, and mark with colour at a height of 60 cm from the foot of the bine plants attacked by viral diseases and any plants of other varieties. The most suitable period for deciding on the purity of a variety and its physical condition ends in Bohemia and Moravia on 5th August and in Slovakia on 25th July. The most auspicious time of day for selection is early in the morning or in the late afternoon, or over the whole day, if the weather is cloudy, i.e. under such an illumination as makes all imperfections in the stand evident. The location of all plants marked for removal and the methods proposed for roguing should be mentioned by the applier in the application which is signed and handed over to Central agricultural committee in Žatec before the official inspection.

The breeder's stand is usually inspected by an official during the growing season after roguing. The total number of other varieties, and of plants attacked by different diseases must not be above the limits required by the Czechoslovak State Standard ČSN No. 46 3790 (see Table 40).

TABLE 40 **Permitted number of other varieties and of diseased plants in authorizated stands**

Diseases and other varieties	Number of non wanted plants in individual levels of propagation			
	S	E	OR	P
Other varieties in mixture[x]	0	0	0	5
Wrapped leaves	1	2	4	5
Necrotic mosaic	1	2	3	5
English mozaic	1	2	3	5
Nettlehead				
Curl[xx]	15	20	45	60

S = superelite;
E = elite;
OR = original;
P = transplantation.
[x] According to this standard, the plants with symptoms of infectious infertility are also taken for other varieties.
[xx] Plants with symptoms of curl will be not destroyed in the stand, but their placement has to be mentioned in the application for authorization procedure.

The stands must not be heavily attacked by bacterial and fungal diseases nor by pests. Plants suffering from curl need not be marked but their position will be mentioned in the application. Marked plants must be removed from the stand after harvest, at the latest by the end of October, i.e. before the second control. At present this work is simplified, because the applicant can destroy selected plants by chemical means (herbicides) immediately during the negative selection (roguing).

If the stand satisfies the requirements, the propagator can then take seed from the stand or use parts of the plants for propagation to produce seed. If plants in the stand were marked for removal, the authorization can be issued only after a second control, when the inspector can confirm their removal. The application for acceptance into the official list of plants grown from seed is governed by the same principles as is that for the mother hop garden.

Production of hop planting material (seeds)

The term 'seed' is used here to describe propagative plant material. The requirements for hop seeds are in the Czechoslovak State Standard ČSN No. 46 3790 – Hop plantation. This standard distinguishes 3 types of hop planting material, known as hop straps, rooted cuttings and container-grown plants. According to the national plan the production of hop seed has to provide every year approximately 3 to 3.5 milion plants for new hop gardens, and for the replacement of missing plants in productive hop gardens. Of this total, rooted cuttings and container-grown plants make up to 90 per cent.

Hop straps

The strap is usually delivered as a modified part of the stem known as 'new wood' which grows from the underground hard part of the hop plant. It is obtained after authorization during the cut by the cutting away of new wood which is then prepared according to certain imposed requirements. Most suitable for the production of straps are young mother hop gardens, where non-traditional agricultural technology is used with the highest possible efficiency. A ridged soil surface in the hop garden encourages the development of long strongly-rooted young wood. It is best to hand cut the hops, because this ensures that the young wood is long enough for its purpose. Under these conditions it is then possible to get a larger number of straps from one hop plant.

The straps are prepared from the thicker basal part of the new wood and if this is of sufficient length then two straps can be obtained from it. The cut surface of the strap must be smooth on both ends. The straps are prepared only from such parts of the new wood as are not wilted, mechanically damaged or attacked by bacterial fungi or pests. Therefore, the straps have to be prepared as soon as they have been cut. The straps must have the characteristics shown in Table 41.

TABLE 41 **Properties of hop strap**

Characteristics	Properties	
	selected quality	standard
Length	110 – 160 mm	70 – 120 mm
Thickness (measured at centre of length)	20 mm	15 mm
Number of nodes (minimum)	2	2
Length of internode (over upper node)	10 – 15 mm	10 – 15 mm
Length of germs	up to 15 mm	up to 25 mm
Minimum weight at planting	30 g	20 g

The mass (weight) of the strap is important. A strap of greater weight has more reserve substances and this allows better development of the above-ground organs of the plant while it is in the plantation and also better growth of the whole system after the plantation. Research work (RYBÁČEK, ŠNOBL, 1969) has shown that if the weight of the strap was 200 per cent greater during its plantation growth there was an increase in the weight of the underground organs by 25 to 75 per cent in the same year. An increase in the weight of the strap also gave a higher seed grade, more vigorous growth and a more even stand as well as an increased setting of hop cones.

Rooted cuttings

A rooted cutting is a modified underground part of one-year old hop plant. The use of rooted cuttings has recently become fairly intensive because it has many advantages over straps. Rooted cuttings are used in newly established hop gardens as well as for the replacement of missing plants in productive hop gardens. Their production is concentrated in a small number of plantations with highly specialized production.

The rooted cutting nurseries have to be sited near to a water source which can provide irrigation. Most suitable are light sandy loam heat-retaining soils free of persistent weeds, with a good structure and containing ample nutrients and humus. If possible only one variety should be planted in any one nursery. Where two or more varieties are planted in one nursery, isolation strips must be made between them in order to avoid the possibility that they might become mingled. The planting of rooted cuttings is still usually done in the nursery in the spring. It has however been shown that planting in the autumn can be advantageous in that it provides 5 to 10 per cent more cuttings of greater weight.

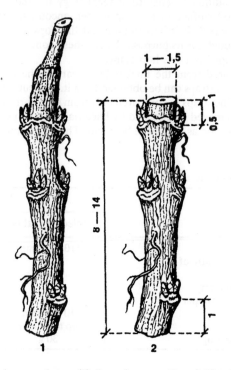

Fig. 70. New wood (1) and prepared strap (2) dimensions: total length 70 to 160 mm, length over upper node minimum 10–15 mm, thickness 15–20 mm.

Before planting the soil of the nursery should be levelled. Planting is either manual into a furrow preciously ploughed or a planting device is used. The distance between the rows will be chosen according to the mechanical equipment to be used for the agrotechnical work. If standard wheeled tractors are to be used, the inter-row width will be that, used in the productive hop gardens, so that the same equipment can be used. If a smaller tractor is used, then the inter-row spacing ranges from 130 to 150 cm. At intervals of six to ten rows a wider spacing should be made in order to allow the passage of a large-scale sprayer. It has been found useful to have two narrow-spaced rows followed by a broad inter-row space suitable for a tractor. The distance between plants in the row depends on the quality of the soil and the intended yield and usually ranges from 15 to 20 cm; distances above 25 cm lower the yield. The planting is done so that the top of the strap is 8 to 10 cm below the soil surface in order to reach moist soil. Experience has shown that superficial planting, to a depth of 5 to 6 cm gave a poor seed grade. This planting procedure was not suitable for the subsequent application of a herbicide drench because the soil could be easily washed away. The planting material must be planted as soon as it is received.

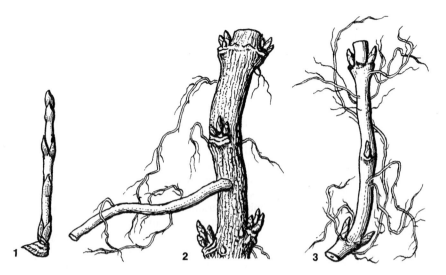

Fig. 71. Planting material used in the preparation of cuttings: 1 – root cutting, 2 – strap, 3 – sucker strap.

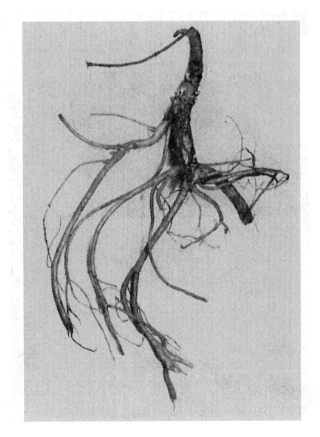

Fig. 72. Rooted-cutting prepared from a sucker.

In the rooted cutting nursery a simple structure is built which facilitates the mechanization of the cultural work. A structure of vertical poles set up within the rows of plants is the most suitable. These poles are linked at the bottom and the top with thick wires with vertically strings twisted between the wires. For training it is possible to use wires 100–120 cm long and 6–7 mm thick perpendicularly inserted in the rows, to which shoots are directed from two or four plants.

In this nursery planting is at soil level or the transplants are heeled in up to 3–4 cm; deep extraploughing is not recommended here. This method produces a better cutting with the required short new wood and with a well developed root system. The young plants need to be protected against diseases and pests, and weeds are controlled by surface cultivations and by herbicides. Herbicides are particularly necessary in the early phases of plant development, when it is impossible to control the weeds by ploughing and cultivating. The correct nutrition and irrigation of the plants is also necessary, especially in the period after cutting, when these factors during July, August and September determine the yield and quality of rooted cuttings.

The lifting of rooted cuttings should start in October after the dying back of the above-ground mass. After the removal of the garden structure and the rooted cutting away of the plant tops, the rooted cutting have to be carefully ploughed out and manually lifted so as to avoid damaging them. Injured roots have to be removed. The rooted cuttings are then classified as to quality, bound together in 20s and transported quickly to the grower or heeled in until planting time. The yield of cuttings from the nursery is between 60 and 70 per cent depending on the variety and quality of the input material, on the climatic and soil conditions,

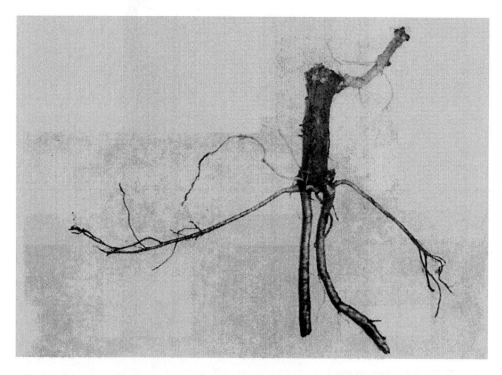

Fig. 73. Rooted-cutting from new wood, with a small amount of rooting.

Fig. 74. Device for excavation of planting holes.

on the application of irrigation, on the level of applied nutrients, on the quality of the agrotechnical procedures and on the effectiveness of protection. Rooted cuttings obtained according to Czechoslovak State Standard ČSN 46 3790 have to comply with the standards shown in Table 42.

TABLE 42 **Properties of hop rooted cutting**

Characters	Values	
	selected quality	standard
Number of roots	minimum 5	minimum 4
Length of roots	120 – 250 mm	120 – 250 mm
Minimum thickness		
(measured 5 cm from root base on a sample		
number of plants)	4 mm	–
Minimum weight of cuttings as delivered	70 g	35 g

Fig. 75. A rooted-cutting nursery with sprayer irrigation.

Container-grown cuttings

Container-grown planting material is one year old hop plants grown and transported in the same receptacle. This is a relatively new type of hop seed, which first came into use in the late 1960s, especially among hop growers in the flat lands.

Nurseries using container-grown planting material have to be in the neigbourhood of a water source, where irrigation is available. The method is particularly suitable for smaller areas which could not be conveniently planted by the more common methods. Black polythene tubing (1 mm thick) is used to make the containers, which are usually 25 cm long and 12 to 15 cm in diameter. Polythene sacks can be used, if they are perforated by cutting their bottom corners.

The preparation of the nursery begins in the winter before the start of spring cultivations. The containers are filled with rooting medium and placed in threes or sixes, upright in rows. Space is left between the rows to allow the sprayer and the irrigation tank to get through. The containers stand on a strip of polythene to prevent the penetration of roots into the soil and thus to facilitate removal of the plants when required. The rooting medium consists of earth and turf or of earth and a humous manuring preparation (Vitahum B) in an approximate ratio of 1:1. The containers are filled from a moving trailer.

Planting must start immediately the planting material is received. Similar material is used as for the production of rooted cuttings; there are straps, one bud straps or straps from suckers. The strap is inserted, with eyes upwards, so that the upper edge of the strap is roughly 7 up to 8 cm below the surface of the medium. Container-grown plants do not need to be stored; they can be planted, even if soil conditions are poor, following bad weather. After planting the land

170

Fig. 76. An unprepared hop rooted-cutting.

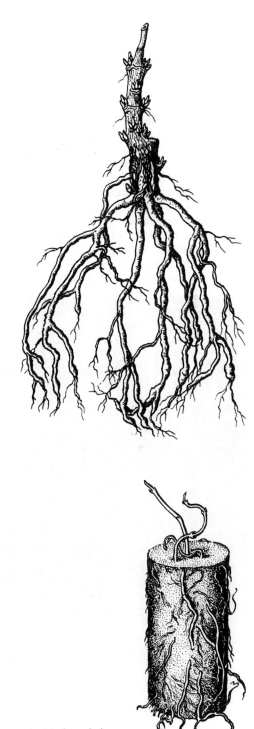

Fig. 77. Container-grown planting material removed with the polythene.

should be ploughed from both sides to earth up the rows. If the straps are ready in the autumn, they can be planted out at that time and give good results. A simple support system is usually built over the rows for the training of a number of shoots.

The plants should receive a multi-component fertilizer and calcium nitrate during the growing season, some given immediately after planting or put into the rooting medium. Protection against diseases and pests as well as the control of weeds in the inter-row spacing is made with chemicals and by agro-technical procedures. The plants should be watered during the season if the amount of natural precipitation is regarded as insufficient. In the autumn the dead above-ground matter is cut off and removed. According to the Czechoslovak State Standard ČSN 46 3790 good container grown planting items should possess a root system that will preserve the original shape of the container after the removal of the cover. The planting material must be free of diseases and pests.

The production of container-grown planting material facilitates highly intensive work. Depending on the inter-row width up to 200 000 such seeds can be cultivated in one hectare. Straps planted in containers grow well, the plants grow vigorously and after transplantation to the hop garden practically all of them will sprout. Moreover they require less attention during the growing season. This method of production however involves greater investment and more work during the setting up of the nursery. But this expense is made up by the harvest in the first year and with earlier full fertility of the hop plants.

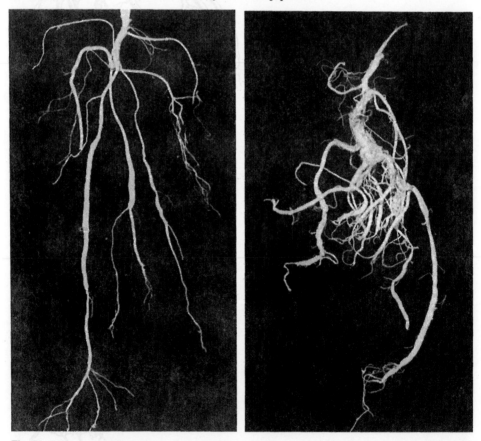

Fig. 78. Different root formation in hop plants produced from straps (1) and from rooted-cuttings (2).

In spite of these advantages and the high economic returns the production of container-grown planting material is limited. This is because production by this method on a large scale needs a lot of manual work which then becomes the limiting factor in its further utilization. There are also problems with the production and transport of the rooting medium. If the palletization of the expedition were to be introduced the need for manual work would be reduced.

Storing and distribution of hop planting material

Hop straps in transit have to be protected against wilting, drying, suffocation and mechanical damage. Therefore, they should be planted as quickly as possible. If this is not practicable they can be stored for a short time in a cool humid atmosphere or heeled into the ground. Rooted cuttings likewise should be planted as quickly as possible, or ridged up in the future hop garden. Straps that are heeled in should be isolated from one another by a thin layer of moist soil and covered by at least a 20 cm layer of soil so as to avoid their early sprouting in the spring with the consequent lowering of their biological value. Before they are inserted into the earth the straps need to have a dry surface but must not be dried out. Recent experience has shown that poor recovery straps after transplanting is usually caused by the wilting and drying out of rooted cuttings. Therefore it is important not to store straps but to plant them directly from the soil into the soil. Container-grown planting material has to be protected against drying and damage to the root cover. Therefore, they too should be taken up from the nursery immediately before they are required for transplanting in the hop-garden.

Straps and rooted cuttings are planting material delivered in sacs, container-grown planting material individually. Hop planting material in transit has to be covered against direct solar radiation. The production of hop planting material in Czechoslovakia is controlled by Chmelařství (i.e. Hop Planting Enterprise) which also controls their distribution. The producer is obliged to deliver hop planting material of the quality required by the Czechoslovak State Standard ČSN 46 3790. The distribution has to be informed when planting material is ready so that it can be handed over to the new planter within 3 days. Inspectors will then examine the quality and health of the consignment which must have no more than 3 per cent of mechanically damaged planting material and no more than 5 per cent of other parameters lower than those in the standard. The distributor gives no warranty for the wilting if this is not claimed during the handingover operation. The distributor is responsible for the variety being that specified up to the first period of full fertility.

THE CREATION OF A HOP GARDEN

The creation of a hop garden is a complicated and expensive venture, which can be fully efficient only when it makes full use of up to date production technology involving its biological, technical and administrative-economical constituents. Therefore it is necessary to consider all of these view-points of all stages of the development, i.e. from the first preparative studies, through the erection of the hop garden structure to the completion of the investment when the garden comes into operation.

The creation of a hop garden involves a complex series of operations, carried out in a limited time, on a compact area. The erection itself starts with the ploughing of the furrows and ends with the decay of above-ground parts of the first years growth. The erection period lasts approximately 1 to 2 years. It cannot be shorter than one year, but may be longer.

Seen from the viewpoint of the financial investment the creation of a hop garden is a broader concept than simply finished garden. BENÍČEK (1977) *distinguished 4 main periods in the creation of a hop garden namely:*

1. design of the layout;
2. preparation of the land;
3. completion of the garden;
4. exploitation of the construction.

According to civil engineering law in Czechoslovakia hop garden is a special agricultural investment within the concept of those minor agricultural building activities with a planned investment of up to 5 million crowns.

Preparing the foundation of a hop garden

The basic work in creating a hop garden starts at least two years before the planting of the crop. Early investigation of the proposed site is essential to confirm that it is suitable for use for hop production. The production of cuttings or container-grown plants takes two years from the time that the propagation material is taken from the mother hop garden. The time required for the preparation of the hop garden and its erection is also two years. This involves not only the design of the layout but also the administrative procedures concerned with the various organizations. Getting the soil into the right conditions for new culture also takes time.

The change of use of agricultural land for the creation of a hop garden

A hop garden occupies farm land for a long period and therefore it is regarded as an independent form of agriculture. The territory occupied by the hop garden is unavailable for use as normal arable land for approximately 20–30 years. For this change of use it is necessary to have the written consent of the local authorities. This can be obtained only after the presentation of a copy of the cadastral map with the area clearly identified.

A hop garden is defined as the area limited by one supporting system. This area depends on the system being used. If it is the traditional system, the area is up to 1 ha, but other systems may be 3–6 ha.

A block of hop gardens unites several individual gardens into one compact area, and its size depends on the number of linked gardens and their dimensions.

A strip of hop gardens joins one or more blocks into one compact complex. If this complex consists of two or more blocks, they need not necessarily occupy an uninterruped area; the area can be interrupted by other agricultural and non-agricultural areas.

The total area of hop gardens owned by one agricultural enterprise includes all of its gardens. This area is increased by the newly created hop gardens from the date of the transfer of farm land into hop garden. The area is likewise decreased if the area of an exhausted hop garden is transferred back to farm land or to other culture.

The area of hop garden culture is regarded as an immovable asset, according to various guide lines. With an accuracy to the nearest 1 m^2 the following dimensions of the culture are recorded:

A – total area of the hop gardens;
B – including
  a) fruitful area;
  b) auxiliary area;
  c) not yet planted;
C – productive area;
D – spring plantings;
E – autumn plantings;
F – increases or decreases of area.

174

The productive area is the area of compact stringing system together with a cultivated border of 75 cm width all round, plus an *auxiliary area* of no more than 100 m². This latter includes roads, paths, spoil heap and other *service areas,* but is regarded as separate, if it is of an area greater than 100 m².

The area is designated as unplanted in the period between the acknowledged transfer of arable land to hop garden and the planting of the hop stand.

The productive area is calculated as the area of spring and autumn planting without regard to the type of planting material (straps or cuttings). Every hop garden has an identification board carrying its number and the sign of the locality mounted on the corner posts of the garden.

A hop garden can be discontinued or transferred to another type of cultivation only with the written consent of the local authority. For the application to be accepted written agreement from the Research Institute in Žatec has also to be obtained.

Preparing for the creation of a hop garden

The preparation procedure is usually concerned with three main tasks:
a) investigation and choice of an area suitable for hop growth;
b) design of the layout;
c) territorial procedure ending with territorial decision.

Their individual duties are closely connected and therefore they must receive attention in the sequence shown.

Investigation and choice of land

The territory in which hops can be grown is naturally much larger than the gardens themselves. This makes it possible to choose those parts most suitable for hop-growing. It is most advantageous to establish a new hop garden in an established hop garden area because the work can be done more economically, the local hop variety can be used and the crop will be cultivated in known, good conditions.

For *the correct setting of a hop garden strip, its size has first to be defined and then its location can be decided within the cadaster of the locality or within the agricultural enterprise perhaps involving other villages. The size of the area covered by the hop-garden strip (P_{CH}) is then calculated according to the following formula:*

$$P_{CH} = a + 1/100 . a . b . (c + 1),$$

where
a = planned area of productive hop gardens;
b = renovation of hop gardens per year (percentage of planned area);
$c + 1$ = number of fodder planting strips as a rotation within the framework of hop gardens; new plantation of hop which gives only small quantity of cones will be added to this number.

The selection of suitable areas takes into consideration such natural conditions as the situation, slope and general shape of the terrain, the local climate and microclimate, soil and hydrological conditions, and the effect on the countryside. Also taken into consideration are such economic factors as the size and accessibility of the land, distance to the harvest centre, distance from an irrigation source, suitability for the erection of the stringing system, and the likelihood that the size will meet the principles of crop hygiene and safety of work.

Only flat or gently undulating terrain with a maximum slope of 6° is suitable for hop-garden strips. The best sites for large scale hop-growing are in valleys and protected dales. If long-term meteorological data for at least ten years are not available, the climatic conditions have to be evaluated from climate maps.

In order to determine the soil quality, geonomic and pedologic maps have to be used together with examinations by specialits. The first job is to take out sondages to a depth of at least 2 m in order to find the level of the water table. A level of 150 to 200 cm is suitable but this level shoul not vary very much during the season or over long periods. The soil profile will show whether the land has supported hops before, which is common in hop-growing regions, or whether it is to be used for this purpose for the first time. This information is very important, because a virgin hop garden requires more cultivation (or fertilization) taking more time than would be the case with repeated creation of a hop garden on the same ground. In both cases the soil is cultivated to a depth of 60 cm, and is heavily treated with organic manure.

Land suitable for hop gardens needs to be of a regular shape and the smallest useful area is 6 ha. The site should be as near as possible to the sources of water and not a great distance from the harvest centre.

The sites of all parts of a hop-garden strip have to appear on the cadastral map. The existing hop gardens as well as those abolished during the previous 5 years have to be shown on the map, because an interval of 5 to 10 years has to elapse between the abolition of one garden and the erection of another one on the same site. *During this period fodder planting sequence is applied to the cleared land as follows:*

1. fodder or sugar beet;
2. maize or other crops suitable for silage;
3. green mixture with under-sown lucerne (alfalfa);
4. lucerne;
5. lucerne.

If a hop garden outside the main strip is abolished, no individual planting procedure is introduced, but this becomes part of the hop strip once it is transferred to hop production.

The selection of the hop-garden strip and of the land for the erection of an individual hop-garden sometimes clashes with public interests and with the interests of various institutions or individuals. These interests need to be considered at a very early stage otherwise the whole process preparation will be hindered. The choice of the hop-garden strip and the use of selected fields is mainly affected by the required protection of the hygiene of residential areas and of potable water sources. Other considerations include non-interference with water supply conduits, road, rail and energy supply equipment.

The zone of sanitary protection is normally at the distance of 100 m from the nearest buildings of a locality. Chemical protection by pesticides applied from the air is forbidden within 200 m. A protection zone of the 1st category is 50 m from water-supply well and 15 m from above-ground water sources.

The protection zone for communications is measured from the centre of the roadway, according to the class of the road:

district road (Ist and IInd class) and local road (Ist class) – 25 m;
district road (IIIrd class) – 18 m;
local road (IInd class) – 15 m.

The protection zone for railways is from the edge of the track. Where the erection of a hop garden would not impair the safety of the road and rail traffic, it is possible to get consent for the reduction of this protective zone.

The protective zone for gas pipelines and equipment are 10 to 50 m. All excavation works including light ploughing are forbidden within 3 m. The production zones for an electricity line are measured from the side cable and are as follows.

Safety zone for very high tension lines:

60 kilovolt (kV) to 110 kV . . . . 15 m;

from 110 kV to 220 kV . . . . 20 m;
from 220 kV to 300 kV . . . . 25 m;
outdoors high tension lines . . . . 10 m.
For other cases of all kinds the protection zone is 1 m.

Elaboration of design

The design of the garden cannot proceed without the results of the investigation and until the choice of the site for the garden, or the block of hop gardens, has been cleared. Only a simplified design sheet is required for the erection of the hop-garden. *This includes the following items:*
1. name, address etc. of the investor and of the erector;
2. calculation of the costs of erection;
3. reasons for the construction;
4. proposed erection procedure;
5. short programme of the work;
6. the dates of the start and completion of the project.
The final consent for the finished design depends on the calculated cost. With a projected investment up to 2 million Czechoslovak crowns, the design may be agreed directly by the investor, but an investment greater than this must be given clearance by a higher authority.

Designing the creation of a hop garden

A hop garden can be regarded as a simple agriculatural venture; therefore design needs to be set out in only a simplified form. The important documents required for the design are as follows: agreed tender conditions included in the "Territorial decision", the most suitable type of hop garden structure and correctly located area. *The documentation of the design includes:*
1. layout of the territory marked on a copy of the cadastral map;
3. plans of the actual layout;
3. an outline of the procedure to be used in the construction for the hop garden;
4. a list of required materials (poles, anchoring bodies, wires) including seeds, fertilizers and other materials for establishing the crop;
5. budget;
6. short economic justification for the project;
7. dates of the start and completion of the construction.
A good design is essential not only for the establishment of the hop garden but also for its long-term success, wherein the necessary annual work is facilitated. The design must include a choice of the type of stringing structure to be used.

Wire-work in hop gardens

Czechoslovak hop culture became famous not only for the high quality of its hops but also for its wirework. This is the so-called "Žatec (or Czech) wirework" which has been adopted in many other countries, where it now serves as the basis of the structures used in hop gardens.
The structure is based on the top construction which consists of the transverse (bearing) wires crossing between the rows of poles and the parallel wires joining the poles within the rows from which hang the stringing rods. The top wirework is supported by the poles which are either internal i.e. perpendicular (Fig. 92) or lateral, i.e. set against the direction of tension (Fig. 79). Lateral poles are fixed by special anchors to ensure the stability of the whole wirework.

177

Fig. 79a. Building the structure of a hop garden.

The spacing of the rows of poles as well as of individual poles in the rows depends on the spacing of the plants. The poles in the traditional wirework, used for short spacing between plants are normally 7.5 to 10.5 m apart producing squares or rectangles (poles in each fifth or seventh row of plants and after each fifth or seventh plant). The height of the structures ranges from 6 to 8 m (usually 7 m). The area of one structure was never more than 7 ha in the past.

Two different basic types of wirework have been used in Czechoslovakia, the hook type and the lug type. The difference lies in the fixing of the wirework at the top and in the anchors of lateral poles. The hook type used a special form of hook to fix the bearing wire to the top of the pole. The lug type had some of the circumference at the top of the pole removed and the bearing wire ropes were looped over the projection (see Fig. 79). Continuing development led to a more frequent usage of the lug type, which is now used exclusively for contemporary structures.

Structures needed to be changed in Czechoslovakia, when hop production became large-scale, because the structures had to meet different requirements:

1. the number of poles which tend to obstruct cultivations had to be reduced to minimum;

2. the height of the structure which was based on the local situation established at 6.5–8 m, now rarely exceeds 7 m;

3. the parallel wires must have the same spacing over the whole of the structure throughout the growing season, i.e. must not be distorted as the growing crop increases in weight;

4. the size of the structure is not limited, but it must be stable and able to withstand summer gales when carrying a full crop. Currently the size is approximately 1 ha but improved modern structures can be 3 to 5 ha. Their stability must be such that individual structures can be linked together to form blocks.

178

Fig. 79b. Erecting a hop pole.

At present two forms of structure are in common usage, the improved *lug type (NOŠ)* and the *large-scale structure (VCHK)*. Their main features are shown in Table 43. In both cases there is now a tendency to replace the wooden poles (usually spruce) with others, e.g. steel, prestressed concrete.

TABLE 43   **Technical data of present day structures**

Data	Normal structure	Large-scale structure
Area (ha)	1	3 – 5
Height (m)	7	7
Spacing of poles (m)	9×6	$9 – 12 – 15 \times 9 – 12$
Poles per ha (number)	168	93 – 118
Weight of wire per ha (tonne)	3.0	3.2 – 3.9
Weight of anchors (kg)	109	50 – 56

Creating and managing block of hop gardens

The optimum area of a single hop garden using the traditional structure is 1 ha. It is rectangular in shape, usually with sides of length ratio 1:2. It is clear from Fig. 80, that the best way to combine individual gardens, is to join their long sides to produce so-called blocks. One block consists of two or more individual gardens. As more gardens are joined together in one block better use is made of the area of the whole strip. If the gardens are precisely linked the rows throughout the block will be correctly aligned making the use of machinery easier and saving time in the manipulation of equipment. If the terrain allows it the rows of hop should be oriented from north to south because it improves the illumination of hop plants in the rectangular spacing. Moreover, if the gardens are joined by their longer sides, their fronts are oriented from west to east, i.e. facing the prevailing winds, which improves their aeration and provides greater resistance of the structures against wind shock.

The aeration of the block is also affected by the orientation and width of the paths. It is recommended that at least 5 m should be allowed for equipment to turn at the ends of rows, but elsewhere pathway of 3.2 to 4 width is sufficient. At the same time the design of a hop garden has to take into account their overall shape, the ratio of their width to length and the required communication network both inside and outside. An important contribution to knowledge of these questions is provided by the study of blocks using the traditional structure in different terrains, as in one Central Bohemian district (Louny).

It was found that *the hop garden blocks had shapes* with ratios of sides ranging from 1:1 to 1:8. Therefore the communication network cannot be fixed particularly, also because the arrangement of block is different on flat land from that in the hills. In flat land it is easier to satisfy the optimum ratio of sides.

The greatest effect on the picture of a block of hop gardens is the length of anchoring systems, their distance from the border wires, and the density of rows of poles as related to the square area of the garden. These factors together with the spacing of plants affect the movements of machinery in the hop gardens.

A survey of the quantitative and qualitative conditions of blocks established in different geographical conditions is given in Table 44.

The spacing of plants affects the whole blocks because of the type of machinery necessary

180

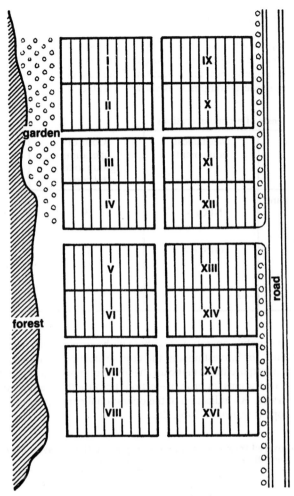

Fig. 80. Arrangement of a block of hop gardens with traditional structure.

TABLE 44   **Dimensions of blocks of hop gardens in different localities**

Locality	Size and shape (%)			
	block size		ratio of lenght to width	
	up to 6 ha	over 6 ha	up to 6:1	over 6:1
Konétopy	86.84	13.16	94.44	5.36
Pnětluky	47.62	52.38	97.32	2.68
Brodec	71.91	28.09	86.85	13.15
Milostín	75.93	24.07	100.00	–
Smolnice	25.98	74.02	100.00	–
Liběšice	72.69	27.31	100.00	–
Average	63.49	36.51	94.44	3.56

181

for their cultivations through other services they require and also by the height to which they grow.

A block of hop gardens has to be arranged and shaped to accommodate the individual operations required for planting and harvesting under local conditions. Most important here are the various mobile operations; an indication of the cost of which under normal growing conditions and using modern equipment is shown in Table 45.

TABLE 45   Direct costs of mobile operations

Group of operations	Mechanical equipment	Costs of operation group to to- tal costs	Note
Autumn ploughing and harvesting	NPCH-4	3	normal condi- tions
Cutting and spring operation	OŘCH-2	37	including ma- nuring
Protection of hop		27	after training 5
Transporting hop to			times sprayed
harvester		33	average distance 1500 m

The most expensive mobile operations are these in the spring which include the cutting, the transportation of bines for harvesting and the crop protection treatments.

The most important requirements in the arrangement of blocks so as to allow the passage of equipment relate to operations for crop protection, because they require the most complicated technological service. An indication of the operational parameters of equipment used for the protection of hops is shown in Table 46.

TABLE 46   Area covered from one full tank by crop protection sprayer. Maximum trip distance with one full tank by crop protection sprayer

Equipment	Volume of tank (1)	Operation width (m)	Dosage (1/ha)	Maximum trip distan- ce (m)
S-029	240	4-5	400-1200	320- 960
S-051	900	2.6-5.2	800-2000	460-3460
Mayers – Taifun	1000	5.2-8.4	800-1500	635-1920
Mayers – Super	1000	5.2-8.4	800-1500	635-1920
Wolf	1250	5.2-8.4	800-1500	790-1490
Kinkelder	1000	2.6-5.2	800-1500	635-3645

The efficiency of mobile operations is mainly affected by the ratio between the time, that it is working, and the time that it is being moved without working, the time required for servicing the equipment and the difference in the speed used in operational and non-operational movements.

The coefficient is used as an indicator of the ratio between working and non-working travel. This coefficient is given by the formula:

$$\varphi = Sp \ (Sp + Sn),$$

where
  Sp = length of working travel in metres;
  Sn = length of non-working travel in metres.
  The optimization of this coefficient increases as it approaches unity ($\varphi = 1$).
  The general relationship between area, side ratio and number of traverses is given in Table 47.

TABLE 47   Time taken to traverse one row as related to block size and shape

Data	Area of the block (ha)								
	3			6.5			26.6		
Ratio of sides	1:1	1:3	1:6	1:1	1:3	1:6	1:1	1:3	1:6
Number of traverses	33	19	14	49	28	20	99	57	40
Total time spent on traverses (min)	16.5	9.5	7.0	24.5	14.0	10.0	4.95	28.5	20.0
Time spent on one traverse (min)		0.5			0.5			0.5	

The table shows that the number of traverses increases with the increasing area of the block but decreases with a closer ratio of sides. Thus in an area nine times larger but with the side ratio altered from 1:1 to 1:6 the total time required for traverses was only about 20 per cent greater.

The size of block and the ratio of its sides affect the coefficient φ, but do not greatly affect the working operations. Coefficient φ increases with increased acreage of the block and rates 0.9 to 3.3 per cent of working path. The growth of coefficient φ is particularly important with regard to the shape of the block. The best shape is one with sides in the ratio 1:6. This coefficient is not linearly related to growing acreage, it decreases above 6 ha and reaches its optimum above 26 ha up.

The communication system of the block is of two types. The first is the roads for the movement of materials and equipment involved in technological services and for the transport of materials from the garden for further processing. The other includes those areas used for the turning of mobile equipment.

The basic communication network follows the normal rectangular system. The best arrangement of the block is that which allows access along the whole length of the front. Within the block, roadways are normally 350 to 400 m apart which is about half of the distance to the service point of the sprayer. The width of these roads should correspond to the maximum width required to finish the trip in one direction, i.e. should be 3.2 m. This width as well as the action radius of used sets is convenient for their turns.

The establishment of the hop garden

Soil preparation required for a hop garden

The preparation of the soil before a hop garden can be established is different from that required for other crops. The hop is a deep-rooting perennial plant, and the crop will remain in this one locality for 20–25 years. Therefore the soil must be thoroughly broken down and its upper layers well mixed with the required mineral fertilizers, before the crop is planted.

After the harvest of the preceeding crop the ground should be *shallowly ploughed* in the summer, and 3–4 weeks later given at least 40 to 50 t per ha of farmyard manure which is immediately ploughed in. This *manure* decomposes relatively quickly during the autumn and its nutrients are freed by the microbial activity in the soil. Before winter the land is given a special, deep plantation ploughing. The manure already partially decomposed is thus mixed with the soil and remains in the upper half of the plantation-ploughed layer, and is available during the first growing season to the root system of the new plants. This deep plantation ploughing loosens the deeper layers of the soil and contributes to development of the root system in these layers. The nutrition of the plants is thereby improved and deeply ploughed soil is better able to retain heat and water. The decomposition of the manure requires aerobic conditions, and so the manure must not be ploughed in the deep plantation ploughing, because this would shift the manure to the bottom of the furrow and its availability to the plant would be considerably limited.

The depth of *the deep plantation ploughing* depends on the mechanical properties of the soil and on the quality of the subsoil. Therefore it is necessary to have information from the use of sondages or soil augers. Generally the ploughed depth is 35 to 60 cm. Land with a thick humus horizon should be plantation ploughed to a depth of 50–60 cm, but where this horizon is thin, it is ploughed only to 35–50 cm. Excessively deep plantation ploughing only mixes less good subsoil with the arable layer and produces poor physical properties, limited microbial activity, reduced soil fertility. On land with a clay subsoil this layer should not be brought upwards but it should be loosened by a subsoiler. After deep plantation ploughing the surface will be levelled with a heavy harrow.

This deep plantation ploughing may bring soil with lower microbial activity and poorer fertility to the surface. Therefore, it is necessary to improve the physical and mechanical properties of this upper layer and to promote its microbial activity. To do this well rotted compost is spread in the autumn and incorporated into the upper layer during the spring. If the subsoil is ploughed out, it can be well fertilized by green manuring. Crops suitable for green manuring can be shown during any time in the growing season and ploughed in when they have produced the maximum of organic matter. The upper layer needs to be enriched by application of fertilizers and certain nutrients can also be added to maintain their reserves.

Green manure can be combined with farmyard manure (FYM) turned into the soil in two doses and to two different depths. The first application (40 to 50 t FYM) is turned-in together with the green manure using a deep harrow or a cultivator and the soil is then given a plantation ploughing to turn the manure deep into the soil. If the hop garden is being established on land where hops have been grown before, then in the autumn a second application of 40 to 50 t FYM will be ploughed into a depth of 25 cm. But if the garden is being prepared on new ground and thus requires long-term fertilization, the second application of FYM will be ploughed in, together with a new stand of green manure as late as the following year, that the land will be ready for an autumn planting of hops.

Before the crop is planted the ground should be well levelled by *dragging* and *harrowing* so as to facilitate planting to the required depth. This is particularly necessary with a mechanical planter. Deeper loosening of the soil surface before planting may be required depending on the soil compactness and planting method. Excessive loosening can, however cause difficulties with the planting holes.

Planting the hop garden

The time of planting

In Middle European hop growing regions, the hop is usually planted in autumn or spring. In the past hop gardens were usually planted in spring, because seedlings were available after the spring cut. A number of research investigations as well as recent practical experience in large-scale production have shown that planting in the autumn has many advantages and since the 1970s this has been increasingly the practice.

Primarily *autumn planting* leads to better growth of plants, sometimes reaching 100 per cent, because the reserve of winter moisture will be used for spring growth. Plants themselves are better established and stronger during the first growing season, they show more vigorous growth and produce a bigger root system, which in turn forms a good base for the next stage development. Moreover yield of cones during first years after planting up to the stage of full fertility of the garden is also higher. It has been shown experimentally that stands from autumn plantings gave, in the first year, 3 to 6times greater yield of cones than hops planted in the spring, though the differences were smaller if cuttings were used. It has to be said however that planting in the autumn is often more difficult, because it is done under difficult meteorological conditions, but its final effect is better, because the stand becomes better established. The autumn here means the period from the second half of October to the end of November, and under good soil and climatic conditions, even during December up to the time or the frosts.

The less successful *spring planting* is mainly due to a lack of moisture during planting, sprouting and initial growth, and is reflected in poorer development, more gaps in the rows and an overall lower quality of the stand. RYBÁČEK (1972) showed that the longer radicles are very often broken during the processing and planting of seed in a spring planting and consequently slows the development and reduces the fertility of plants. Furthermore the apical dominance and physiological properties of such damaged seedlings are changed. Another disadvantage of spring planting is the pressure of other work during the spring, which can delay the planting operation. Experience has shown it to be necessary to limit spring planting and totally to exclude late spring planting. Thus spring planting should be done only in emergencies when, for various reasons, it was not possible to do it in the autumn. If the planting in spring is unavoidable then irrigation of the garden is very positive in its effect on better development and growth.

Depending on soil and climatic conditions, the period of spring planting starts at the beginning of April. Provided the plants are available so early the best time is the first half of April, because this allows the best possible use of the spring moisture. The work can then go on up to the first half of May, but after this time the deficiency of soil moisture becomes significant. Sometimes, even planting in the second half of April is not successful, the risk in the late spring planting is generally higher for straps than for cuttings and container-grown plants.

Type of seeds used for planting

The type of seed material used has a significant effect on the quality and fertility of the hop stand during the first years after planting. The traditional method of establishing a new hop garden by the use of straps is not suitable for present-day large-scale production. Due to their low content of reserve substances and their low mass (weight), straps have a lower sprouting rate than cuttings and container-grown plants under less favourable soil and climatic conditions. The stands they produce are often less robust, show many gaps and reach full fertility relatively late. The number of gaps in such stands usually increases because after the first growing season some of the plants remain underdeveloped and damaged during cultivation. This gappiness significantly lowers the yield and requires rapid correction. The planting of straps is also more costly because it requires more attention during their early years.

Cuttings and *container-grown material* are, therefore more advantageous for the planting of a new hop garden. Their advantages have been demonstrated by both research work and practical experience, in the field. They form stands which are fully developed and with few gaps they do not require any additional planting. The sprouting rate and development after the autumn planting of cuttings or container-grown material is 100 per cent, whereas for autumn planted straps the figure is approximately 85 to 100 per cent and for spring planting it is 60 to 90 per cent. A selection of the straps before planting excludes those that are underdeveloped and this contributes to the vigour of the stand. Full fertility can be reached one year earlier, and this gives valuable extra hop production. If straps are used for planting the hop garden, full fertility will be reached under normal operational conditions in approximately the fourth year, if cuttings are used, it could be in the third year and with container-grown material it may be as early as the second year. If cuttings are used then in the first year it is possible to reach a yield of 500kg dry hop per ha whereas straps will produce only 100 kg per ha. Cuttings and container-grown material show high vigour after planting and produce higher yields every year until stand reaches full fertility. Later practically no differences were observed between straps and other planting material. ŠNOBL (1969, 1974) found that, when cuttings were used for planting, their rooting system develops much more vigorously than that of straps. Four year old plants, developed from cuttings produced underground mass (weight) 10 per cent greater than those grown from straps.

The effect of different types of planting material on the yield of green hop is shown in Table 48; this includes data gathered by SACHL and KOPECKÝ (1970–1973).

TABLE 48 **Effect of type of seed material on yield of green hops**

Type of seed material	Mean yield of green hops (kg per plant)			
	1st growing season	2nd growing season	3rd growing season	4th growing season
Seedlings	0.08	0.60	2.60	3.54
Cuttings	0.77	2.44	3.20	3.65
Container-grown plants	1.62	3.22	3.85	3.22

Container grown plants have a better developed rooting system due to the better nutrient supply from the substrate during the first years after planting out and this is reflected in the higher yield of cones. This substrate supply becomes exhausted by the fourth year when the yield suffers a slight depression. Therefore, it is necessary, during subsequent years to improve the nutrient supply to the more vigorous rooting system in order to achieve the expected yields in later years.

In the future, hop gardens should be planted only with cuttings and container-grown plants; hop straps should be used only in nurseries. This will ensure the earlier fertility and higher yield of newly planted stands.

The spacing of straps

Inter-row and inter-plant spacing is developed according to the machinery used in the agricultural operation in hop gardens. The traditional spacing, at 150 × 150 cm used for small-scale production, is not suitable for the larger machines used in modern large-scale production. Therefore the spacing was changed to 260 × 110 cm and 280 × 100 cm and combined with the training. Recently the spacing between rows has been changed from 280 to 300 cm, because it is more convenient for modern equipment. This also means that the percentage of destroyed and damaged plants is reduced and the efficiency of the machinery used for inter-row treatments increases by 10 per cent with the reduction in the number of traverses per unit area of garden. Likewise the introduction of the 300 × 100 cm spacing reduces the cost of establishing the stand.

Technology of planting

This technology has to ensure a precise spacing at 300 × 100 cm without any deviation of the plants from the rows. Similarly, the depth of planting must be constant, because this is vital for the mechanized cut. Furthermore the technology of planting has to provide ideal conditions for sprouting, so that plants must be put into moist soil, and the soil must be ridged-up to the plants.

New hop gardens are planted either manually or mechanically.

The traditional manual planting into holes is still very popular, because it satisfies most of the requirements. But it is very costly, both of time and labour. It consists of measuring and marking the planting, of digging the holes and the manual placement of the hop seeds and their covering with soil. This work requires much labour and often the work is not finished by the required time. Quicker methods for the digging of holes came into use in the past in order to make the work of hop planting easier. Where the soil is light, it is possible to use shallow furrows ploughed out in advance, to put the seeds there and to cover them by hand. This method is difficult with heavy soils.

If the hop garden is established in ridges or with the use of cuttings, then the depth of planting is governed by the mechanical properties of the soil. The traditional depth used in the past will be modified according to local requirements. The final depth of the plants reached some years after their planting depends on the depth of the cut. Generally, the planting should be shallower (10–12 cm) in heavy soils and deeper (12–13 cm) in lighter soils to take account of the upper layers possibly drying out during the growing season. The depth of planting is the perpendicular distance from the upper edge of the seed to the surface of the soil. The seeds are planted with eyes upwards in order not to damage them. Good soil tilth should be obtained and the planting holes should be dug immediately before planting in order to avoid the adjacent soil drying out. The planting is so arranged that the seeds are always put to the same side of the hole and covered with well-broken, loose and moist soil which is well firmed in. This firmed layer is later lightly covered with the rest of the soil. This looser layer ensures the rapid warming of the soil as well as the easy penetration of the shoots to the surface. The upper part of the hole has to be dish-shaped to collect rainwater. Only one piece of seed is planted to form one hop plant.

Because the labour requirement for manual plantation is high, attention has been given to two aspects of *the mechanization of planting, digging the holes and the complete machine planting.*

For hole digging the equipment is connected to the auxiliary hydraulic take-off at right side of a tractor (in Czechoslovakia Zetor 3011) where it can be seen and controlled from the driver's cab (Fig. 81). The hole-digging auger is 20 cm in diameter and 1 m long, and the hole depth is controlled by an adjustable foot. When used for planting-up a new garden, two augers are fitted, but when used for filling gaps in an established garden only one is used. Holes over an area of 0.7–0.9 ha can be dug with this equipment in one working shift.

Mechanical planting makes use of a semi-automatic machine which simultaneously plants two rows of seedlings or cuttings. The plants are dropped, by hand, into the holes made by the digging machine. The machine consists of two side-by-side seeding units with a three point connection to the tractor. Each seeding unit has one seated operator and consists of a frame, cutting knife, soil-opening share, seeding drum, axle with wheels, covering discs and seat. The operator puts the plant into the inserting segment of seeding disc which as it rotates takes the plant to the bottom of furrow made by soil-opening share. After the insertion into the furrow the seed is covered with fine soil by small share and then the whole is covered by covering

Fig. 81. Detail of the equipment used for digging planting holes.

discs. The seeding discs are driven from the rear wheel of the tractor by chain transmission. The depth of planting can be set from 5 to 15 cm below the surface. The required number of plants in the row is set up by an ajduster on the drive chain and by changing the number of segments in the seeding disc. The position of seeds in the soil may be horizontal, vertical or inclined, with eyes towards the surface. The machine can plant 2–2.5 ha per shift.

Since the 1960s *hops have been planted in some areas on the flat*. Here, container-grown seeds are used packed into shaping vessels. The vessel hinders the growth of lateral suckers and confines the underground parts to the required shape. Thus, the shape of the horizontal roots has, therefore, to be carried out by others. The rooting system develops very rigorously below the shaping vessel, whilst the vertical roots grow normally. This shaping vessel limits the area covered by shoots growing out after the cut and so makes training easy. This type of seeding makes certain operations much simpler and therefore, economically advantageous.

Unlike the traditional plantation in hills the container-grown seed is made on the flat and planting is relatively shallow. The upper edge of the shaping vessel is usually placed 5–7 cm below the surface. Planting is either manual in previously dug holes, with eyes upwards; a semiautomatic seeding machine modified for this type of operation is used. In any case the creation of a new hop garden on the flat is more expensive.

Treatment of a hop garden during its first year

New hop gardens have to be treated with care during their first year. All work on them must help in the development of the above-ground parts of the plants because this is closely related to the development of the underground parts.

188

The garden has to be kept *weed-free* during the vegetative period as weeds can become established very quickly in this unshaded area. The weeds are removed by normal cultivations or by herbicides and at the same time the soil is loosened and aerated. Chemical weed control has now become an integrated part of agrotechnical interventions, but preference should be given to control by cultivations because hop plants, especially when young, are liable to be easily damaged by herbicides. Herbicides should be used in the spring, when weeds cannot be ploughed in, and are usually applied before the first hop shoots appear. No further cultivation of the hop garden is necessary until the additional ploughing. Subsequently, the growing weeds have to be removed by cultivations with or without the use of herbicides.

Growing shoots must never be allowed to creep over the ground without any support. It is also incorrect to train them on poles or wires only 5–7 mm thick because the hop plants overgrow them and thus their total growth becomes limited. Rapid growth of the plant and a successful harvest of cones can be assured only if the structure is erected before the seeding operation, because the training wires can then be connected to them and used for the training of shoots. If an autumn planting is made of selected cuttings in conditions favourable for growth, and irrigation is applied, then it is possible to connect 4–6 shoots to 2 training structures arranged in a V-shape. If spring planting is done using cuttings or straps, then only one training wire may be attached and all of the selected shoots are trained to it. For training all the usual and current principles have to be obeyed. The plants do not all grow at the same rate; the shoots grow individually so that after 2–3 weeks it is usually necessary to train the slower growers. Subsequently, the training involves those shoots that have fallen or wandered.

After the completion of training an *additional ploughing of the plants* should be made. For this process, two lateral discs are suitable, because they remove weeds both between and within the rows, ploughing them in and loosening the soil.

In a newly planted hop garden an *intensive artificial fertilizer dose* should be delivered. This should be well balanced but have plenty of nitrogen, because this helps the development of above-ground parts, and ample potassium because this encourages the development of the root system. An important part of the treatment of hop gardens is an efficient protection of the plants against diseases and pests. If the garden is near to a water source, the plants should be irrigated according to local climatic conditions to encourage good spring growth and, in the summer, to promote the development of the under-ground as well as the above-ground parts.

During the first year the newly-created hop garden will give a *small yield of so-called virgin hop,* in quantity varying according to the weather, type of seed material used, and various other factors. The cones are often of even worse quality and mature later than those in old hop gardens. Therefore, they have to be harvested later, usually after the harvest of older gardens, and mainly by hand because normal decapitation cannot be applied. The bines remaining after harvest have to be cut down after the transfer of assimilates from the above-ground parts to those underground. This benefits the fertility in succeeding years.

Other cultivations in new hop gardens, made during the autumn, after the bines have gone, are similar to those made in productive gardens.

TECHNOLOGY OF THE CULTIVATION OF PRODUCTIVE STANDS

Improvement of productive hop gardens

Productive hop gardens often suffer by not having a full complement of hop plants. *As well as natural biological reductions, there are other reasons for plant losses; these can be classified as follows:*

1. general level of planting of the foundation of the garden;
2. effects of machinery and the level of agrotechnical interventions;
3. effects of applied chemicals.

To reach the optimum number of well-grown hop plants, the right number must be planted when the garden is created. Furthermore it is necessary to check the sprouting rate and to replace missing plants in order to get a full stand in the next growing season.

The greatest contribution to the losses of plants is made by the machines used for the various treatments in the hop garden. Of particular importance is the loosening of plants in the autumn, the mechanized cut and various treatments during the growing season. Then there is the usage of herbicides and desiccating preparations. Hence it is evident that it is necessary to check, continuously the condition of plants growing in the productive areas. According to SACHL, KOPECKÝ and ŠTRANC (1977) the annual loss is 0.5 per cent of the plants. Theoretically, therefore, a twenty years old hop garden should have lost a maximum 10 per cent of its plants provided that the total was 100 per cent at the start. Reality usually differs from theory and it is essential that plant numbers are recorded, from time to time, as a basis for the maintenance of a full complement.

The Research Institute of Hop Culture in Žatec has elaborated methods for listing and improving hop gardens.

The inventory of stands is made in the following way: on each hectare of hop-garden three checking areas are marked on a diagonal. Each area contains 100 plants so that when the number of missing plants is noted, the mean of the results is the percentage of missing plants.

Fig. 82. Semi-automatic machine planting container-grown cuttings.

If the annual loss of 0.5 per cent is assumed then numbers are made up at the following times:
1. soon after planting;
2. in the 7th to the 8th year;
3. in the 15th year.

If this schedule is used then the percentage of missing plants should not exceed 5 per cent. After the investory of stands has been made the areas are classified according to the number of missing plants and those hop gardens are improved first where the loss is greater than 5 per cent.

In order to reduce such losses, it is recommended that the following principles be adhered to during the foundation of the hop garden:
1. the best time for planting a new garden is the autumn, up to 30th November, because at this time the soil condition, particularly the moisture level, is at its best;
2. cuttings of the best quality, i.e. the "selected quality" of the Czechoslovak State Standard ČSN 46 3790 should be used; container-grown cuttings are the most suitable because they guarantee the best plant growth;
3. the cuttings must be planted so that their top is at least 2 cm below the level of the subsequent cut;
4. after planting it is necessary to cover the cuttings, to hang the wires and to train at least 4 shoots; for the additional ploughing a disc-type plough is used at a low speed in order not to cover the sprouting shoots.

Autumn treatment of hop gardens

The autumn treatment consists of cleaning the area after the harvest and preparing the soil for winter.

Cleaning the hop gardens

After mechanical *harvesting* there is a residue of plant material left on the soil surface. Theoretically, this being the part of the shoot which does not bear cones should be at least 1.5 m long. However, harvest decapitation is usually somewhat lower, at about 1 m. After *manual harvest,* which is best on a new hop garden, the whole of the above-ground part of the plant is left behind from whence the assimilates are transferred in this after-harvest period for storage underground. Therefore, the clearing of the above-ground parts of the plants is delayed for so long as assimilation continues. This depends on climatic conditions but usually lasts until the second half of October. Excluded from this procedure are years with severe downy mildew because then it is necessary to finish the cleaning before the winter spores of *Pseudoperenospora humuli* pass into the soil, i.e. before the second half of September.

The remnants or whole bines are cut off by hand or by means of special equipment. In either case 20–30 cm of above-ground old hop plants should be left. These remnants mark the place of individual plants in succeeding operations and show where there are missing plants.

When the above-ground parts of the hop shoots have been removed, then the garden *is harrowed twice,* in diagonal directions, to remove hop strings and weeds.

For this purpose a special four-part harrow is used consisting of a fixed central part and two lateral collapsible arms for better manipulation and transport. The central part of the harrow beam is connected to the tractor with three point linkage. The working width is 4 m, operation speed 5–7 km.p.h. and a transport speed of 10–15 km.p.h.; 8 ha per shift can be harrowed. This harrow levels and loosens the surface of ground and turns in the artificial fertilizers before the autumn ploughing. After the removal of the side segments it can be used to loosen the soil between the plants during the growing season.

Fig. 83. Tined harrow used in the hop-garden.

Autumn processing of the soil

Because the traditional planting method, using the ridged system, has been largely superseded by different modern planting methods, observations on the autumn processing of soil are now changing. But the new variants are not yet fully elaborated so there is now a variety of observations to be made. In the soil and climatic conditions specific to Middle Europe, only shallow ploughing and deep loosening are applied.

Autumn ploughing between the rows remains as the main autumn treatment of hop gardens. For the turning in of FYM green manure or in the struggle with damaging weeds, there is nothing better than autumn ploughing, because no substitutes have given equally good results. This applies not only to the turning of organic or artificial fertilizers, but also the reversal of the upper layer to the required depth, which thus brings to the surface the materials carried down by heavy rain and by irrigation. This ploughing helps the development of the soil microflora and the rooting system of the hop plant. Therefore, it is vital to maintain the structure of the soil particularly in view of the current high level of mechanization with the consequent large number of passages of middle and heavy machinery used in crop protection and harvest. The depth of this ploughing usually ranges from 15 to 20 cm depending on the soil properties, the age of the hop garden and its condition after previous treatments.

An attached multiple-furrow plough is used for autumn ploughing. This plough is suitable for any width of spacing as used in the various hop-growing regions. It can plough 8 ha per shift, or approximately 120–160 ha in a season, working speed 3–4 km p.h., transport speed 10–15 km p.h. It can operate in a maximum inclination of hop garden of 6°. The requirements of this plough are that its dimensions must be suitable for use in hop rows, giving good

192

Fig. 84. Plough used for the autumn ploughing in hop gardens.

loosening of the soil to a depth of 15–20 cm. It must satisfactorily incorporate organic and artificial fertilizers. If it is used for loosening the unploughed strip down the centre of an inter-row, then its use will be combined with that of other equipment.

Although ploughing cannot be substituted and has many advantages, other treatments can be given during the autumn. Nowadays, therefore, nobody insists that autumn ploughing must be done every year.

Heavier soils with a high percentage of clay particles (over 45 per cent) tend to settle and the extent of this depends on the water regime of the soil. It is also affected by the technology used in the treatment of the garden and by the machinery actually employed. The result is the consolidation of the layer below the topsoil forming so-called pan. This situation exists not only in hop gardens, and is described by the variable used for the general pedological classification of soils, as reduced unit mass (weight), porosity etc. As a result of this 'pan' formation other variations of the basic treatment have to be used in some hop gardens e.g. deep loosening or pan-breaking.

Subsoil ploughing of inter-rows

This process was developed in the Research Institute of Hop Culture in Žatec. A special subsoiler is used for this work. Its construction is shown in pictures 85 and 86. The working capacity ranges from 3 to 4.5 ha per shift and 100 ha per season; working speed 2 to 4 km p.h., transportation speed 15 km.p.h. The machine is used in the autumn after the hop gardens have been cleared and when the soil is in the right condition (its moisture content should not exceed 20 per cent).

193

Fig. 85. Subsoiling plough used for pan breaking.

Loosening involves an intensive and uniform breaking up of the whole profile between the rows resulting in improved physical properties of lower layers of the topsoil and the subsoil. It is possible to break down the soil to 60 cm depending on the mechanical structure and physical properties of the profile. The machine has to pass down the centre between the rows so that the distance between the loosened soil and the longitudinal axis of the rows of hops is not less than 70 cm (isolation distance).

Deep loosening once in 3–5 years positively benefits the yield. As well as the advantages already mentioned it also improves the water and air regime of the soil which is important in water economy. Deep ploughing stimulates the root system and promotes root regeneration and consequently gives better plant nutrition. However, this intervention cannot be repeated at short intervals, because even partial injury of the root system causes a depression of yield.

Shallow surface loosening came into use after the introduction of minimal cultivation. This involves a fairly intensive loosening and levelling of the surface to make possible the smooth working of the machines used for the spring cut. High work productivity and low consumption of energy are its particular advantages. Working at such a small depth then has no great positive effect on the subsoil horizon and its physical condition can be expected. The depth of loosening is usually 20 to 25 cm. If the shallow loosening is done in two directions at right

194

Fig. 86. Subsoiling plough fitted with fertilizer drills.

angles then the depth in the same direction as that of the rows can only be to the upper level of the old wood. This method, however, does considerable damage to the old wood, particularly that of young plants put in as replacements.

Shallow loosening cannot replace deep ploughing. It can be justified only in exceptional cases and extreme conditions such as extraordinarily unfavourable climatic conditions. It can be a valuable additional treatment during the autumn if it is, for example, necessary to loosen the soil before normal ploughing because the surface has become compacted.

Spring treatment of the hop stands

Cutting and associated spring operations

The aim of the cutting operation is to remove excess new wood and some of the lateral suckers from the underground old wood. These increments of new wood have to be cut off underground, therefore the cutting operation and the preparation for it require considerable physical fitness and experience among those doing the work. This was the main reason why 'the operation was frequently limited or even totally omitted, as mentioned many times in the literature.

TOMEŠ (1891) quoted the experience of HERMANN, who planted seedlings only just below the soil surface and lying almost horizontally. Some of the runners produced were trained to the wire and others with their side suckers, could be easily destroyed with a hoe as they were not deep below the surface. HERMANN explained that in order to ensure its survival when the young wood with the largest shoots was removed, the plant was induced to sprout many lateral suckers. This method was criticised when it was found that many runners developed when no cut applied and the removal of these runners caused losses equal to those resulting from normal cutting. Some cultivators who adapted the method of noncutting, found that their yields gradually decreased.

Attempts to exclude or at least to limit the cut in those foreign regions where conditions of growth are different and there is a different composition of varieties are described by ZASUCHYN (1909), PYŽOV (1964) and others.

Neither in Middle Europe nor in other hop-growing regions was the cut totally excluded although in the U.S.A. and elsewhere the operation has been reduced or replaced with less complicated interventions as e.g. the destruction of the first sprouted runners by mechanical (knocking, harrowing) and chemical means and by fire.

Recently consideration has been given to the mechanization of this operation, which had been regarded as almost impossible if applied the traditional individual cutting methods in view of the diversity of the underground parts. However, once it was shown that the hop could be cut in one horizontal level by recently designed cutters passing uninterruptedly along showed that the technology of mechanized cutting the hop plants was feasible.

According to the numerous literature references (MÜNTZ, 1827; TOMEŠ, 1821; GROSS, 1904; BLATTNÝ, OSVALD, NOVÁK, 1949: LINKE, REBL, 1950; POKORNÝ, 1957; RYBÁČEK, KONEČNÝ, 1966) and to the results of modern research work, *the importance of cutting may be condensed to the following points:*

1. The cut governs the period of sprouting of the runners, the length of the growing season of the above-ground parts and, consequently, governs the development of organs. This control of the sprouting period and the optimization of the term for training the runners is valuable in Middle European hop-growing regions. It makes possible the regulation of the time schedule governing the growth of the crop to match the time schedule of length and intensity of illumination, as well as of other climatic factors, which affect the morphology of the plant, and thus regulates the number and size of hop cones economically required.

2. Annually repeated cut substantially limits the lateral growth of underground parts of the hop plant because it removes or shortens the underground suckers.

3. Cutting limits the area producing runners and their training to the wires consequently needs less manual work.

4. The plants maintain their original spacing.

5. The permanent underground old wood remains at the required depth.

The date of hop cutting and its effect on organogenesis

Cutting may be done in the spring or in the autumn.

The autumn cut can be from the second half of October to the beginning of winter. In the past, before the introduction of suitable machinery, the autumn cut was used in those farms with large hop gardens so as to reduce the spring working peak because all plants had to be cut in a relatively short time interval. Even today some agricultural enterprises follow this schedule of work, because it allows them to start early in the spring with the fixing of bines to wires. Furthermore, if the production of hop seeds for autumn planting or for container-grown nurseries is required, the hop gardens are usually cut in the autumn.

From the biological point of view, however, the autumn cut is unsatisfactory, because it causes very early sprouting runners if the spring is early and warm, and later in the season cones become overripe on prematurely ageing plants.

The spring cut is the most suitable from the cultivation point of view and is also the most commonly used. The cutting period lasts from lst to 25th April according to local weather conditions, particularly the temperature which thus allows for the training of runners to wires to be done between 5th and 25th May.

Control of sprouting runners by covering them with soil

The date when autumn cut plants (or even early spring cut) sprout runners can be regulated by the cut underground parts of plants being covered with a layer of the soil the depth of which can be different. The effects of the different thickness of this cover or the date of sprouting and on the growth of hop plants is shown in Fig. 87 based on results gathered from 1969 to 1972.

The diagram shows, that a soil covering of more than 10 cm retards sprouting more than is required. Thus, with covering layers of 15 and 20 cm, the sprouting was so retarded that the runners did not reach the 50 cm length required for training until after 26th May, i.e. after the agrotechnical limit. These retarded runners grew relatively quickly at the beginning of June, but the subsequent growth of their organs was much affected.

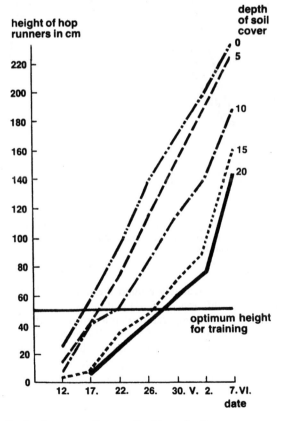

Fig. 87. Effect of depth of soil cover on the growth of hop plants.

The speed of sprouting and growing is directly correlated with the thermal regime of the soil immediately around the old wood, as is shown in Fig. 88.

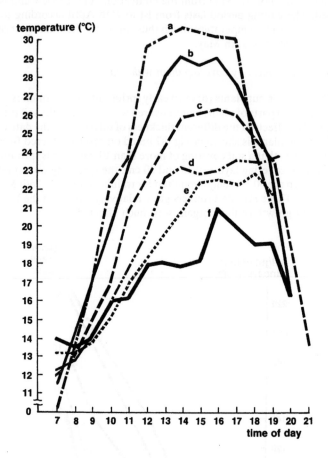

Fig. 88. Temperature of soil near old wood, under soil layers of different thickness, at different times of the day: a – air temperature, b – no cover, c – the cover 5 cm thick, d – the cover 10 cm thick, e – the cover 15 cm thick, f – the cover 20 cm thick.

The temperature of soil immediately around the old wood between 7 a.m. and 8 a.m. ranged from 11.4 to 14 °C (difference 2.6 °C) and between 1 p.m. and 2 p.m. from 21.1 to 29 °C (difference 7.9 °C). The temperature depth profile, with regard to the position of sprouting eye, also affects the uniformity of the developing runners. The eyes on manually cut old wood lying at different depths underground and, therefore, covered by soil of different depths develop less uniformly than the eyes of plants cut by machine at one horizontal level and covered by a compact layer of soil of constant thickness (see Fig. 89).

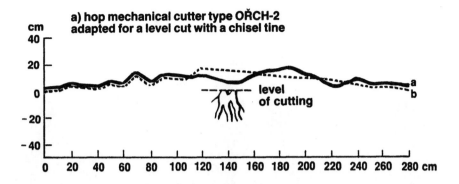

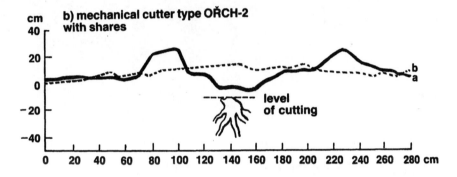

Fig. 89. Transverse profile of hop garden soil before (a) and after (b) cutting with two types of cutting equipment.

Technology of the mechanized cut

A mechanized cut differs from the manual cut because all of the plants are cut at the same fixed depth, the operation is uninterrupted and the machine is continuously in work. The machines cut away the one year old wood and the lateral suckers by a flat cut just above the wooden part of the old wood. The cutting device can be swung in order to get round hop poles. The cutting machines made in Czechoslovakia have a rotary cutting disc. Some foreign cutters have counter-rotating discs with sharp edges, so that the operation is similar to shearing.

Morphology of underground parts changed by repeated mechanized cut

Changes after repeated mechanized cutting involve an increased number and length of lateral rootstocks. An original mean number of 15.8 laterals of mean length 5.5 cm increased after six years of mechanical cutting to 31.2 and 6.3 cm, respectively. Moreover the old wood grew in size from an original circular shape of diameter 19.08 cm to an oval, with its longer axis in the direction of the operation, of 31 cm and the shorter axis of 22 cm.

The difference between the level of cut and the top of the original old wood is greater with a mechanical cutter, so that the remaining twigs of the new wood are on average up to 3.82 cm in length. This is unavoidable because the cutter damages plant tissues below the actual cut surface and the damaged tissue later decays. The disc rotating more rapidly and equipped with

199

Fig. 90. Two hop cutters – on the left the type for inter-pylon rows, on the right the type for normal rows.

Fig. 91. Cutter, showing the cutting disc.

200

more teeth reduces the amount of damage but it is always greater than that caused by gentler manual cutting. Therefore the uncut length must be such that the damage affecting the tissues will not reach the original old wood.

Mechanical cutting may damage the underground organs but this does not necessarily involve any reduction of the vigour and longevity of the hop plant. The separation of the old wood into a number of independent sectors (NEČIPORČUK, 1955) guarantees the continued growth of the plant even, when one part of the old wood is separated by being torn off vertically. The hop plant is not directly endangered if parts of the old wood are separated horizontally, as for example, by the lowering of the level of the cutter. The eyes and buds are spread along the whole vertical axis of the old wood, so that new runners will be produced, after a delay even when part of old wood has been cut away. As the plant ages the old wood becomes an increasingly complicated body more resistant to the effects of mechanical damage. However, each injury to the underground organs, particularly if repeated, increases the risk of infection and the consequent death of the whole plant:

SACHL (1965, 1966, 1968) *found that mechanical cutting causes:*

mean annual reduction of the stand equal by 0.3 per cent resulting from mechanical damage or complete destruction,

damage to the tissues of the old wood at the cut surface due to:

crushing produced directly by the teeth of the disc and by pressure transferred to neighbouring tissues (to a depth of 0.9 cm);

splitting produced by the forward pressure of the cutting disc (to a depth of 12.5 mm).

After cutting there are problems with the fixing of the wires in soil near to the old wood, which must be done quickly. The actual positions of the plants are not very clear so that the wires are inserted into the soil in approximate positions, and may have to be moved later. The sprouting runners are usually more scattered and so their training can cause problems particularly in older hop gardens.

Basic procedures for mechanical cutting in productive hop gardens

If the planned productive life and yield of the garden is to be reached, *the following principles concerning the function of the cutters and the protection of underground parts of the hop plants must be respected:*

1. it is necessary to prepare the ground for the cut and to level the soil surface;

2. the working speed of the cutter must be between 2.4 and 3.7 km p.h., i.e. third or fourth reduced gear of the tractor must be used;

3. there must be 40 teeth on the disc and they must be sharp enough to cut into new wood;

4. the depth of the cut (measured from the levelled surface of the ground) must be at 5 to 10 cm in a productive hop garden;

5. the depth of tissue crushed in seed material must be a maximum of 5 mm;

6. the thickness of the soil cover may be 4.5 to 5.5 cm according to the time of year of the operation;

7. the setting of the cut above the old wood may be up to 1 cm.

Cutting in seed-producing plantations

Mechanized cutting can also be applied to hop gardens where hop planting material is produced as well as to those for the production of hop cones. The yield of hop seeds is greater where traditional manual cutting is used but mechanical cutters are being increasingly used because manual cutting is very labour-intensive. If cutters are used then the maximum yield of seed requires that the level of the cut is close to the boundary between the old and the new wood and the cut is clean, giving a smooth surface even at the cost of a very limited working speed of the cutter. Therefore, cutting discs of greater diameter with more teeth are used here, because of the higher cutting speed and shallow engagement thus involved.

If mechanical cutting is used in seed production, it is necessary to use a working procedure which avoids damage to the new wood. This involves good preparation of the ground, lower working speed, discs with a higher number of teeth and the correct adjustment of working depth.

The collection and treatment of seed should follow immediately after the cut in order to avoid the material drying out and losing its vitality. Therefore, the area to be cut is that which can be dealt with at once.

Perspectives of cutting

The mechanization of cutting gave generally beneficial results and made possible the further development of large-scale technology. This development goes on though it would be a great advantage if cutting could be avoided. However, the properties of the Middle European hop are such that if cutting is not used then some other method of controlling the sprouting period needs to be discovered. Likewise, a suitable means of governing the development of underground parts would be needed using methods advantageous for economy and operation.

Fixing the training wires

Hop plants are cultivated on supports which they climb to shape the stand, to make best use of solar radiation and which gives sufficient space for the production of cones in favourable microclimatic conditions.

TOMEŠ (1891) showed that cultivated hop had for long been trained on individual wooden poles of 5 to 8 m according to variety. Hops were trained on wires in Middle Europe for the first time in 1850.

After World War I hemp ropes were used for stringing (MOHL, 1924), these were carried over the wirework by means of different weights and were then fixed with a tie. The suspension of the ropes and wires was simple, they either moved up and down the training pole or were fixed at its end.

In the 1930s the ropes were replaced by cheaper steel wire of approximately 1 mm diameter. At the beginning this wire was fixed by means of a special device but later by a simple bar to a hanging hook (ZÁZVORKA, ZÍMA, 1956).

Core rope, used for training, was described in England by BURGESS (1964), who described the method of fixing by means of a bar to hooks made from thick wire fixed on the top of the wirework. In the United States (BROOKS, HORNER, LIKENS, 1961) the core ropes are previously cut to the length corresponding to the height of the construction, then from a transportable platform they are manually tied directly to the wires. The height of the working platform is that which gives access to the wires in the rows. A similar method of fixing the strings is used in Yugoslavia.

In various countries, including the Federal Republic of Germany, Belgium and France, steel wire is mainly used but recently polypropylene string has come into widespread use in all hop-growing countries.

In Middle European hop-growing countries steel wires are used, annealed so as to make them malleable and easy fixing. The important technical properties of such wires are shown in Table 49.

According to the data of ZEISIG, KREITMEIER (1970) and KOHLMANN, HEINDL (1975) the wires used in West Germany have a diameter of 1.1 to 1.4 mm. In order to avoid the collapse of suspended hop plants during the vegetative period, the wires must have

Fig. 92. Suspending the strings.

TABLE 49 **Properties of steel wires**

Diameter (mm)	Weight per ha (kg)	Tensile strength (kilogram-force, kgf)	Length per kg (m)
1.00	350	24.9 – 33.9	162.1
1.06	410	24.9 – 33.9	144.3
1.12	450	27.5 – 37.3	129.3
1.25	540	34.4 – 46.6	103.8
	hooks 65		

a tensile strength of 35 kilogram force (kgf), because this decreases due to corrosion by 17 to 20 per cent. New wire should therefore, have a tensile strength of 42 to 44 kgf. The strength of the wires should correspond to the mean mass (weight) of hop plants and be 39 kgf for light varieties and 47 kgf for heavy ones.

Strings made of steel wire

It is not advantageous to use the steel wire in hop gardens intended for large-scale production. Its suspension by hand is a difficult and slow operation, which every year causes a peak demand for labour. This can be managed only with the help of seasonal workers and requires complicated organization as well as extensive and expensive social and sanitary facilities.

Steel wire used for the suspension of hops splits and cracks due to the curling movement of the bines. Its strength is decreased by corrosion the unfavourable effects of which are intensified by the compounds used in crop protection sprays. Summer storms often cause the collapse of hop plants and to re-suspend them is troublesome. The suspension hooks cause difficulties during the harvest, because they damage the picking equipment. The hooks also make it difficult to use the remnants of hop plants as forage because they contain also strings and wires. It is therefore necessary to equip the picking machines with a means of catching these undesirable sundries.

These negative features of steel hooks are removed if the steel wires are replaced by plastic strings and their means of suspension is changed.

Strings made of plastic materials

Polypropylene strings are used which are specially treated to accommodate Middle European, and specially Czechoslovak varieties, which have a limited climbing ability and limited adhesion to the string.
The properties of a plastic string:
a) resistance to weathering, especially to solar radiation, and a strength ranging from 24 to 28 kgf at the time of harvest;
b) rough surface and sufficient flexibility to encourage the adherence of bines to the string during the whole vegetative period;
c) modification of stringing to make mechanical harvesting possible;
d) not harmful if used with the after-harvest remnants of hop plants as fodder;
e) minimum stretching;
f) resistance against the effects of chemicals and microorganisms in the soil;
g) economically advantageous.
Tests of suitably modified polypropylene strings showed their advantages. The number of collapsed plants during the vegetative period was reduced, the need to use suspension hooks

disappeared along with all the troubles associated with their use. Thus, the way to mechanization and automation of the suspending and harvesting of hop plants was open.

Tests of polypropylene strings in Czechoslovakia demonstrated their complete applicability. *Further development will be concentrated on the following points:*
a) to increase the roughness of the surface;
b) to reduce variations in their strength;
c) to obtain a better length to weight ratio;
d) to achieve total decomposition of the remnants.

Their practical application also depends on an efficient and automated suspension on the wirework and fixation in the earth.

Methods of achieving suspension

The most common method, applied in hop-growing regions all over the world, involves *the suspension of strings* using a *travelling platform*. The strings, in bundles, are usually cut in advance to a length corresponding to the height of wirework and the individual strings are manually fixed to the top of the construction. The fixed strings are then pulled out of the bundles as the platform moves forward. The most recent platforms, used in the United States and in Yugoslavia have a special mechanism which draws the string down and fixes its lower end in the earth by means of a small anchor or by ploughing it in (PELIKÁN, 1976).

Czechoslovak research institutions are working on a design for *the automatic suspension of strings*. An automatic suspender must produce a reliable method of fixing the strings, both to the top of wirework and in the earth, in the neighbourhood of the old wood (maximum distance 15 cm away in the row and 5 cm to the side). Suspended strings have to be sufficiently tight. The fixing at the top must be designed so that the plants can be easily taken off during the harvest and to allow for the after-harvest usage of bines. The strings must be so fixed on the wires in the row that they do not interfere with the suspension in the following season. Another important requirement for the platform is that it must be able to get past the cross lines of the wirework.

This operation seems to be central to the studies of suspension problems in Czechoslovak research institutions.

Training the runners

Training of the runners is the remaining operation not yet mechanized. As yet there has been no method devised to replace this manual work by a mechanical operation. This operation is indeed decisive, because it has an important impact on the yield and quality of cones.

Date of training

The effect of the date of training of the hop runners on the yield-building elements, is shown in Fig. 93. The data were gathered from a long-term experiment at Stekník (Bohemia) from 1970 to 1973. The highest yield of fresh cones from one plant (2.05 kg) was given by plants trained 12th May, whilst the lowest (1.26 kg) was from plants trained as late as 1st June. Late training reduced the yield by 38.5 per cent. However, early training had a negative impact on the yield, which was 10.3 per cent lower from plants trained on 4th May. As training was delayed so the mean length of harvested cones tended to decrease but their density of setting increased. In other respects the colour of cones was poorest with the earliest training but the content of α-bitter acids was highest with longest vegetative period. All this clearly shows that the date of training principally affects the yield of cones and their quality.

Present-day cultural treatments are now clearly based on the optimum date of training. Their biological characteristics show that young hop runners are much affected by low temperatures. Thus in a long cold period there is only slow regeneration of the above-ground organs accompanied by physiological symptoms. The effects are evident throughout the vegetative period and result in a decrease in the yield of cones. The climatic conditions in Middle Europe, therefore indicate that the best date for training the runners is after 10th May. At this time the runners should be of a length suitable for training, i.e. an average of 60 cm. In those years when the cold period is long and pronounced, it is best not to train long runners (i.e. not to select the longest) but to train those which sprout later and which have been less exposed to the effects of cold. Once suitable conditions occur it is necessary to start training at once, otherwise the yield will be reduced.

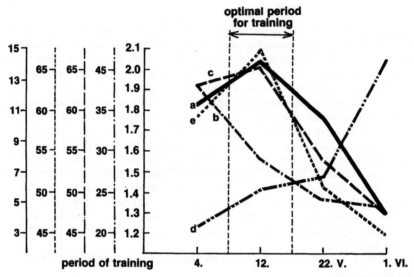

Fig. 93. Effect of time of starting training on the structure of the hop plant and on the yield of cones: a – yield (in kg) of fresh hop per plant, b – length of cones in mm, c – number of shoots, d – density of setting (number of cones per 10 cm of shoot), e – mean length of shoots (in cm).

The schedule of other operations, especially the cut is based on the requirements for training. The sprouting and development period of runners also differs according to the depth of old wood, and the chemicals used for weed control. Hence the time of the cut is critical, because it has the major effect on the sprouting period and the development of shoots ready for training at the right time.

Under working conditions the training period is usually long, because once conditions are suitable for growth the runners quickly grow into each other and the work becomes difficult and slow.

In terms of organization therefore, it is necessary to catch the right moment for the start of training and at this time to proceed as quickly as possible while the stand is in the optimum condition.

Methods of training the runners

Hop runners being dextrorotatory climb around the support in a clockwise direction. Only healthy runners should be selected and they must not be mechanically damaged. They should arise from the centre of the sprouting runners and the best of these arise from the centre of the old wood. The height of the runners after training should be uniform and therefore it is not the longest runners that are selected.

Several methods of training can be distinguished.

In the past *perpendicular training* was the most common. After the turn over to wide spacing the arrangement was for two strings to leave each plant; the strings are inclined from the perpendicular to form a V-shape. These strings often sag if plant growth is vigorous, so the two strings are so joined at 120 to 150 cm above ground level to form the shape of the letter Y.

Inclined training has also been tested, and for this the strings slope upwards from the stand. This gives an increased length of bine and therefore, the height of the wirework can be lower. This variant was tested by SACHL (1975), who found that this form of training with an operating height of 6 m has the same results as a perpendicular height of 7 m. This in comparison, however, is valid only when the yield is compared of plants with no growth defects, i.e. with a full number of bines growing up to the height of the wirework. Under Middle European conditions it is often found that this training procedure results in excessively high numbers of runners becoming detached from the strings. The Middle European varieties, unlike those grown elsewhere in the world are significantly limited in their ability to climb. It was found in a test series that from the ideal number of two bines trained on one string 1.58 grew up to the full height of the wirework with perpendicular training, 1.42 where bines were sloped by 80° and only 1.18 where bines were sloped by 70°. Sloped training also caused a greater number of unattacked vegetative apices (10.1 to 25.4 per cent more than with perpendicular training) and there are further problems with sagging. The strings should be inclined towards the four cardinal points, but this means that at the peripheries the structural pieces are close together, so the construction itself is costly and there are serious problems with the harvesting of the bines.

The training procedure is not only of single runners. With the aim of achieving maximum yield from the required number of runners it was usual in the past, to keep a number of reserve runners, which were removed at the second training, at about 120 cm providing that all the first choice runners were progressing normally. Nowadays however many hop producers do not keep reserves because of the labour required for their removal.

Training also includes the so-called *corrective training of sloped runners,* which has to be done regularly until the runners reach the wirework. Above this height the bines have to be trained from travelling platforms.

Number of trained runners

One of the important factors governing yield is the number of trained runners per ha. Under Middle European conditions the optimal number of bines is 13 000 to 14 000 per ha. This condition can be reached with correct number of plants per ha and correct ratio of runners trained from each plant.

Therefore, the question arises as to how many runners can be maintained by one plant and how to avoid a depression of yield if too many are trained. According to the experience of Middle European growers, one hop plant is able to nurture six runners and more with very favourable irrigation and nutrition. Therefore, from the practical point of view *four bines (runners) trained from one plant* give a reliable yield factor and at the same time they make it possible to train more runners as a replacements for any missing plants.

Table 50 shows how many plants have to be trained with 6 runners depending on the percentage of missing runners in a 300 × 100 cm spacing. The upper limit of trained runners is used as a basis because during the vegetative period a certain percentage of runners will not reach the full height of the construction and will thus not give a normal yield.

Another important question is the number of runners to be trained on one string. This

TABLE 50   Numbers of plants with six trained runners and the effect of different
percentages of missing

Missing plants (%)	Real number		Number of missing runners (full stand 14 000 runners)	Number of plants with 6 trained runners
	plants	runners		
0	3330	13 320	680	340
1	3297	13 188	812	406
2	3263	13 052	948	474
3	3230	12 920	1080	540
4	3197	12 788	1212	606
5	3164	12 656	1344	672
6	3130	12 520	1480	740
7	3097	12 388	1612	806
8	3064	12 256	1744	872
9	3030	12 120	1880	940
10	2997	11 988	2012	1006

question was considered even before the introduction of structures. TOMEŠ (1891) recommended that two runners and exceptionally three should be trained on one pole. This question was discussed by OSVALD (1947) and his unequivocal conclusion was that the optimum number of runners on one string is two.

Self-trained runners

The Hop Research Institute in Žatec examined the question of self-training as a means of simplifying training. SACHL, KOPECKÝ and PAZDÝRKOVÁ (1975) found numbers of self-trained runners as shown in Table 51.

In the training period 43 runners grew from one old wood, in the traditional furrowed cultivation in an area of 1500 cm² (38.75 × 38.75 cm). After a low horizontal cut, repeated over several years, 69 runners grew on the area of 2000 cm² (44.75 × 44.75 cm) with the result that the runners of neighbouring plants were mingled at the time of training.

Studies of the gyrations of hop runners were made by KOPECKÝ (1972) (see Fig. 94), who found that the diameter of the apical movement depends broadly on the height of the runner. If a runner encounters a support it is very likely to fix itself to it.

The self-training of hop runners offers the possibility of simplifying the operation if it is combined with the use of desiccants to remove superfluous bines.

TABLE 51   Numbers of self-trained runners on strings

Year	Number of strings (%)			
	no trained runners	one trained runner	2-3 trained runners	4 or more trained runners
1972	26.8	15.6	50.0	7.6
1973	17.3	12.6	61.8	8.3
1974	20.9	24.7	43.0	11.1
Average	21.7	17.7	51.6	9.0

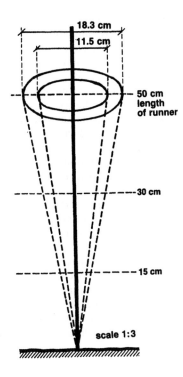

Fig. 94. Diagram to show the gyration zone of a growing shoot apex.

Summer cultivation of hop gardens

This cultivation can be seen as two operations, one the cultivation of the soil, and the other the treatment of the stand.

Cultivation of the soil

This operation involves two basic procedures: the loosening of inter-row soil, and earthing-up. Each of these is done at different times with regard to the local pedological and hydrological conditions and to the degree of weed infestation.

Loosening the soil between the rows has, according to the traditional concepts of agricultural technology several beneficial features; it reduces water loss from the soil and improves its aeration, promotes microbial activity and consequently improves the mineralization of organic manures as well as making nutrients more readily available, and it mechanically removes young and established weeds. At the same time it destroys roots growing in the surface layers and consequently activates those lower down where there is a greater resource of moisture and it promotes the regeneration of the upper part of root system. Contrary to the earlier practice of very frequent loosening of the soil, the contemporary method relies on a limited cultivation of the soil.

The first loosening between the rows follows immediately after the end of training. This aerates the upper layer of the soil which will have become flattened during training. With the newly developed loosening equipment, equipped with blades similar to coulters it is possible to work down to a depth of 12 cm. This cultivation is usually followed by ridging-up with bilateral disc ploughs. If the newly-developed loosening device with lateral coulters is used for this work, both operations can be done simultaneously.

In the second half of the vegetative period, i.e. during the flowering and cone-forming period these cultivations have to be reduced to a minimum, if the soil is very humous or sandy. During this period it is possible to loosen the soil only down 8 cm so as not to destroy the rooting system. During this period the root-hair system grows in the surface layers, where its activity contributes substantially to the production, growth and quality of hop cones, and therefore it must not be damaged.

The number and depth of loosening operations as well as their timing must depend on the moisture content and moisture-holding capacity of the soil, which is affected by its mechanical composition and its tendency to subsidence.

Heavier soils with reduced water-holding capacity in uneven terrain must be loosened before each irrigation in order to make uptake possible. Soils, with a tendency to subsidence and to compaction, have to be loosened immediately after irrigation when they are in a suitable condition for loosening.

The next operation, e a r t h i n g - u p , brings soil from both sides to the rows of hop plants. Once the under-ground part of the plant is covered by soil it becomes histologically changed young wood, from which the new hop seedlings can be formed. The depth of soil cover provided by the earthing-up, therefore, controls the yield of seed material.

The first shallow ploughing should be done immediately after training the runners, i.e. from 25th May to the latest by 15th June, according to the local conditions.

Fig. 95. Loosening the soil between the rows of hops.

This ploughing improves the nutrition of the hop plant through the summer root-hairs which appear afresh every year on the covered parts of bines. This improved nutrition should not be overestimated, because this part of the rooting system, produced on those parts of the bines covered by the earthed-up soil represents only a small part of the whole complicated rooting system of the hop plant.

In those hop gardens where no planting material is produced (productive hop gardens) only one earthing-up (up to 10 cm deep) is sufficient. However, if weeds are not controlled by herbicides, during the summer shallow ploughing has to be repeated.

In seed-producing hop-gardens, the plants are usually earthed-up two or three times with a deeper layer of soil, so that its final depth reaches 15 to 20 cm.

Heavy mechanized traffic in the inter-rows causes the destruction of soil structure and a deterioration in physical properties to increase caking and penetration rate of water. PAZDÝREK (1972) found that the soil density, which serves as the indicator of consolidation, increased at the surface by 21 per cent after one passage of the tractor and after three passages it increased to 32 per cent. The damaging effect of the pressure of wheels decreases with increasing soil depth but it is evident from the above-mentioned data, that the amount of tractor movement during the summer has to be reduced to a minimum. Moreover, attention should be paid to the deep loosening of the soil in the autumn in order to avoid permanent deterioration of soil properties, because the soil structure throughout the whole active profile cannot be restored by loosening during the vegetative period.

Tending of crop

The limited climbing ability of Czechoslovak hop varieties causes the vegetative apices to come away from the strings during the period of their intensive growth, i.e. in time interval between 25th May and the end of June. This is mainly caused by the winds which occur at this time.

The retraining of dislodged vegetative apices is an important and unavoidable manual operation during this period. The productivity of a hop garden can be estimated from the percentage of bines which grow to the height of the construction as a proportion of the theorical number. This parameter is satisfied if more than 85 per cent of bines grow up to the height of the wirework.

This evaluation of the number of bines reaching the wirework is used as the basis of the reward salary system for the workers in that particular hop stand. This rate at this time indicates the likely future harvest.

As the crop grows so the danger grows that some strings will break due to winds or summer storms and that the bines will fall. Such events cause losses in the quantity and quality of cones because some of them are bruised, damaged and soiled. If a fallen plant is not immediately suspended, the cones may be attacked by downy mildew or by hemp flea beetles. To raise fallen plants, wooden bars and simple lifting devices are used and the bine is re-suspended with hooks to the top of the wirework.

Irrigation of hop

The irrigation of the hop has become increasingly important in recent times. It is very important in the region of Žatec, which is one of the driest regions in Europe.

The annual precipitation there is approximately 500 mm (in wet years it may be up to 700 mm, but in dry years it can be as little as 100 mm). In the town of Žatec the precipitation reaches only 445.6 mm on average. The probability of a dry year ranges from 5 to 50 per cent.

The long-term annual mean temperature is approximately 8 °C and for the vegetative

Fig. 96. Re-training of dislodged runners.

Fig. 97. Re-training of displaced runners being done from a mobile platform.

212

period it ranges between 12 and 14 °C. Saturating deficit ranges from 2.5 up to 3.5 Lang's rain factor goes from 60 up to 70. Hops can be grown under these conditions because most of the rainfall is during the summer, when the plants have most need of moisture. However, this need is greater than the quantity available from precipitation and therefore must be supplemented from underground water. Additional irrigation is also valuable as a means of supplying moisture to the biologically most active surface layers of the soil. Here the root-hair system is active and here, therefore, the amount of nutrient available for uptake during the different developmental phases of the hop plant will control plant growth and determine the ultimate yield of hops.

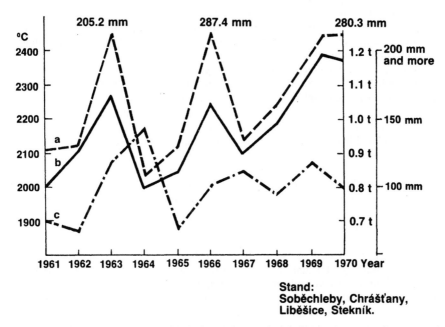

Fig. 98. Hop yield as related to precipitation and temperature during the growing season (i.e. from lst May to 31st August): a – total precipitation, b – yield of hops, obtained in four localities, c – total temperature.

Only a small part of the hop-growing regions has irrigation systems, and so the bigger part cannot be irrigated. The capacity of the presently available water sources is sufficient for approximately 2800 hectares of hop gardens. It is, therefore, the main task of the hop growers and the various technical departments in these areas to provide irrigation to the hop gardens where there are economically available water sources. Irrigation of other hop gardens is possible only where there are ponds.

The effect of irrigation on the yield and quality of hop was demonstrated in experiments by SACHL (1971) and has been confirmed in practice by BAMBÁSEK (1972), PALKOSTKA (1972) and others.

Technology of irrigation

The traditional methods, used in the 1920s and 1930s, were based mostly on furrow irrigation. More recently sprinklers have come into use.

Furrow irrigation

Irrigation from inter-row furrows can be used in those hop gardens, where sprinklers are not available provided that the pedology and the state of the terrain is suitable. The losses resulting from the use of this method are usually bigger than from sprinklers, but its advantage is that little equipment is required, because water comes directly via stopcocks from distribution pipe lines. *It has greater disadvantages, however, and these can be summarized as follows:*

1. excessive consumption of irrigation water in pedologically unbalanced conditions, resulting from imbitition;

2. the distribution of water is not uniform;

3. considerable deterioration in soil structure;

4. the amount of irrigation to each plant cannot be fully regulated and measured;

5. water accumulates in the lower parts of the hop garden, and the resultant muddy areas impede mechanical equipment;

The principles to be adhered to in furrow irrigation:

1. the length of the furrows is usually 100–150 m up to 250 m in good pedological and terrain conditions, but less than 100 m in light, absorbant soils;

2. the spacing of furrows is governed by the width between the rows;

3. the slope of the ground should not exceed 15 per cent;

4. the calculated irrigation dose should, in practice, be increased by 20 to 50 per cent, to replace the loss of water due to this type of irrigation.

Repeated furrow irrigation increases the danger that the natural soil fertility will be degradated by leaching. Because of the pedological conditions and the relief of the terrain, most of the hop gardens in Czechoslovakia are irrigated by sprinklers.

Sprinkler irrigation

The majority of hop gardens use sprinkler irrigation although its requirement for technical equipment is much greater than that for furrow irrigation. Sprinkler irrigation is suitable for hilly terrain, for absorptive soils and for sloping ground.

O v e r h e a d   i r r i g a t i o n   o f   h o p   g a r d e n s   h a s   s e v e r a l   v a r i a n t s :

1. *Irrigation by individual sprinklers placed above the structure.* This is a relatively simple system consisting of a portable pipe line with a take-off valve and a perpendicular extension tube below each sprinkler. A long-distance sprinkler is fixed by a bayonet connector into the top of this tube; positioning of the tube and its suspension on the wirework is done manually.

2. Irrigation line 100-ZLCH designed by Sigma (Olomouc, Moravia) in cooperation with the Hop Research Institute in Žatec. This line has three main parts: fixed, portable and connecting. The most important fixed part is the aluminium distribution pipeline, mounted permanently on pylons. This pipeline has hydrants joined by a portable plastic hose to the sprinklers, fixed by bayonet union to the erectable stand pipe. The sprinkler is lifted to the level of the structure by means of a pulley and wire rope.

The sprinkler can be lifted above the wirework by the rotation of the perpendicular stand pipe fixed to the pylon. The rotation of the sprinkler about the central point facilitates its replacement and repair.

This design removes the need for portable pipelines which require much labour but at the same time it involves a considerable requirement for the maintenance of a fixed distribution system and its reliable operation during a relatively short irrigation period.

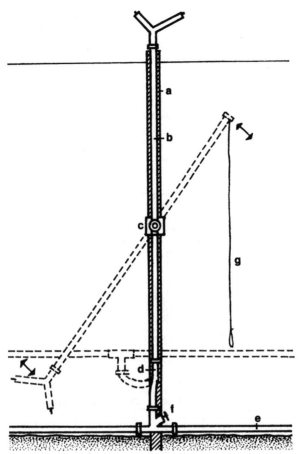

Fig. 99. Sprinkler arrangement for the irrigation of hop gardens: a – fixed pylon, b – perpendicular extension tube, c – rotation point for raising and lowering sprinkler, d – rubber hose, e – permanent pipeline laid on the earth or fixed on pylons, f – branch pipe with valve, g – wire rope for raising and lowering sprinkler, full lines – sprinkler in working position, dashed lines-sprinkler in servicing position.

Zonal sprinkling

The main feature of this irrigation method is the replacement of a fixed or portable distribution system by one plastic hose of a length of 200–300 m and an automatic winding drum.

The hose which has a sprinkler at its end is unrolled from the automatic drum with the help of a tractor which takes it to the most distant point to be irrigated. The water is turned on and the sprinkler is then slowly drawn back towards the drum. Thus a strip of the hop garden is sprinkled of a width up to 27 m and a length of up to 300 m.

The sprinkler for zonal sprinkling must have many orifices because the water jets penetrate the stand at a height of 2–3 m and must not cause damage to the plants.

The *essential advantage* of this system is its high efficiency and simplicity as well as its lower cost, in comparison with the traditional method using pipelines. This equipment, however, can be fully exploited only in large gardens where its action radius can be fully used.

One of its *disadvantages* is excessive irrigation. The mean irrigation requirement for argillaceous soils is 15 mm per hour, but zonal sprinkling provides 2–6 times more than this. It provides 29.5 mm per hour if rotary sprinklers are used, and as much as 91.4 mm per hour, if stable side jets are used.

This greater level of irrigation demands a higher imbitition rate of the soils, particularly on heavy soils, where the maximum precipitation usually ranges from 6 to 12 mm per hour.

In some cases surface run-off will result in the accumulation of water in the lower parts of the garden. The distribution of irrigation is thus uneven and the resultant swampy areas make subsequent cultivations and the application of protectants difficult. With an increasing slope of the hop garden, the difficulties with irrigation increase. Zonal irrigation is most efficient but only on fairly level hop gardens with soil possessing a high imbibition rate. This rate can be improved if the soil is loosened before the start of irrigation.

A zonal sprinkler with a water delivery rate of 340 l per hour, with a water under pressure of

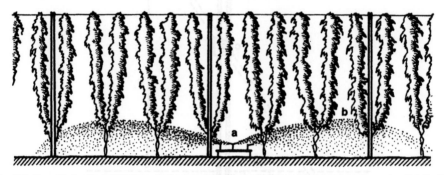

Fig. 100. Zonal irrigation of the hop garden: a – travelling sprinkler with many jets, b – water distribution.

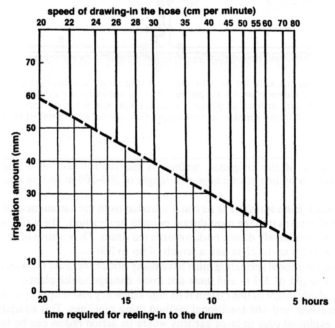

Fig. 101. Zonal irrigation. Relationship between rate of reeling-in of sprinkler and amount of irrigation. Inlet pressure of water 0.7 MPaG and width of irrigation zone 27 m.

0.3 MPa and a travelling speed of the sprinkler of 35 cm per minute, would provide 30 mm irrigation over 1.1 ha per 24 h operation.

Combination of fixed sprinklers with reeled hose

The disadvantages of the two irrigation methods can, to a certain extent, be limited by a combination of sprinklers fixed above the structure with the zonal, reeled-hose, water supply system.

Zonal sprinklers with higher efficiency resulting from the use of longer polythene tubes (up to 300 m) of greater diameter (90 mm) and an increased rate of flow (600 and 900 l per minute), make possible the use of jets having a larger radius of throw (35 to 39 m). The irrigation thus provided is more homogeneous, it covers a larger area from each traverse and the amount of irrigation water applied does not exceed the maximum requisite values.

The reeled hose is positioned equidistantly between the two neighbouring rows of fixed sprinklers. Wherever the sprinklers are in use the drum is sheltered so that the overhead irrigation water does not come into touch with the driving mechanism thus making it possible to provide foliar nutrients from the overhead equipment. The production of zonal sprinklers of greater efficiency began in Czechoslovakia at the end of the 1970s and their use in large-area hop gardens has now been verified.

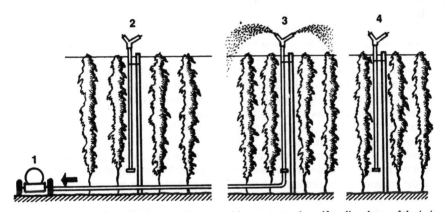

Fig. 102. Combination of fixed sprinkler with a reeled-hose system: 1 – self reeling drum of the irrigation line, 2 – fixed sprinkler next to be used, 3 – fixed sprinkler in use, 4 – fixed sprinkler after use.

Underground irrigation

The use of underground irrigation has not yet spread far into agricultural production, but its advantages suggest that it could be used in hop gardens. The network of this system consists of plastic tubes with effluent holes protected against the ingress of roots. The tubes are laid at a depth of 40 cm so as to allow subsequent ploughing. Water slowly escapes from the holes, to irrigate the soil along the whole length of the pipeline and to moisten the root zone. If irrigation is excessive, then water begins to be lost to lower layers. DONEY (1975) *gave the main advantages of underground irrigation as follows:*

1. the operation is automatic and the working costs are lower;
2. mechanized agricultural production processes can continue because there are no pipes or other irrigation equipment on the surface;
3. the structure, aeration, and thermal properties of the active layer of soil are improved;

4. higher yield with low water consumption.

The application of this method of irrigation to large scale hop production is under research in Czechoslovakia.

Determination of the irrigation schedule

It has been demonstrated that the growth and development of hop plant occupy two periods, when the moisture content of the upper layers of soil to a depth of 60 cm should not fall below the minimum needed for the absorption of nutrients.

These two periods are in two critical phases of hop growth, one when the shoot comes into flower (in the first half of July), and the second during the growth of cones (in the second half of July and first half of August).

The growth and development of the assimilative vegetative organs is almost complete and the production of fructiferous organs reaches a climax in the *period prior to blossoming*. The amount of irrigation applied during this period is a decisive factor in the size of the plant and the number of inflorescences.

The whole productive apparatus of the hop plant is engaged in the production of organs or propagation during *the cone-producing period*. The growth conditions during this period determine the weight and size of cones and thus govern the quality and size of the harvest. The consumption of moisture during this period is at its maximum.

The requirement for irrigation of the hop garden depends on the timing of the main growing periods and on the development of the plant as well as on the amount of precipitation before irrigation and the pedology and general meteorology of the locality.

The important growing periods and the developmental intervals suitable for irrigation in a productive hop garden are the following:

1. shoot development up to first blossoming, usually from 1st July to 15th July;
2. production of cones, usually from 16th July to 8th August.

In seed-producing hop gardens the most suitable intervals are the following:

1. sprouting, usually from 15th May to 15th June;
2. first growth of shoots, usually 15 th June to 30th June;
3. shooting, usually from 1st July to 15th July;
4. blossoming and production of cones, usually from 16th July to 8th August.

When the reserves of underground water are low and the quantity of precipitation does not satisfy the moisture requirement, a hop-producing garden can be supplied with required amount of irrigation during June.

Total amount of required moisture (V_c) as additional irrigation in hop-growing regions is expressed as that amount of added moisture, required to satisfy the absorption of moisture during the vegetative period of the hop from the biologically most active part of rhizosphere down to 60 cm. This total amounts to 2500 m^3 water per one hectare. The vegetative period of the hop is taken as the period from 1st May to 31st August.

Optimal irrigation doses, established for hop (m^3 per ha):

loamy sand . . . . 260 m^3 i.e. 26 mm,
sandy loam . . . . 300 m^3 i.e. 30 mm.
loam. . . . . 400 m^3 i.e. 40 mm,
clay loam . . . . 460 m^3 i.e. 46 mm.

These directory irrigation doses should not be exceeded so as to avoid losses of water by leakage and other disadvantages with excess water. When the total irrigation volume required to make up the deficit seriously exceeds these directory doses it should be applied in repeated doses.

The amount of irrigation, as related to the depth of effective soil moisture is given in Table 52 (from DERCO et al., 1975).

The determination of exact basic data for the management of an irrigation schedule has

TABLE 52 **Amounts of irrigation to be applied, related to soil quality and depth of effective soil moisture**

Depth of effective soil moisture	Amount of irrigation water (mm)		
	light soil	medium soil	heavy soil
20 cm	15	18	13
40 cm	30	36	26
60 cm	45	54	39
80 cm	60	72	52

recently come under investigation. One way towards the management of the soil-moisture condition of the hop garden is to arrange the irrigation schedule according to the total water emission (evapo-transpiration), the soil moisture content and the physiological requirements of the plants by means of the biological curves of moisture consumption developed by PASÁK and SLÁMA (1975).

A different method is to manage the irrigation according to the changes of energy within the soil based on the temperature and precipitation conditions in the previous most fruitful years, i.e. by a graphical and analytical method (KUDRNA, SLAVÍK, 1971).

The aim of research in this field is to produce a forecasting system for the need for moisture, i.e. a system which indicates the date of the first irrigation dose and the succeeding irrigation schedule, which could be valid for all those Czechoslovak hop-growing regions with similar ecological conditions.

A survey of technological variants of hop production

Since World War II there have been important changes in hop growing and in the management of hop gardens. New methods of management suitable for large-scale production, have been studied, based on new scientific knowledge. This process has taken many years. A comparison of the traditional, current and prospective management of hop gardens is shown in Table 53.

The soil treatment for individual variants is given in Figs 103–105.

The mechanization must correspond to the adapted system of culture. *The following machines and equipment are required for the present methods and for the management systems likely to be in use up to 1990:*

1. planting of hops: semi-automatic planter for the foundation of new hop gardens, dibbler for general and for replacement planting, semiautomatic planter for the foundation of nurseries for cuttings;

2. autumn treatment of hop gardens: heavy, toothed harrow, subsoil plough, single-share, double-sided plough, multi-share plough,

3. spring treatment of hop gardens: heavy, toothed harrow, hop cutter clear, device for the mechanical removal of prematurely sprouted runners, device for the suspension of strings,

4. treatment during the vegetative period: disc harrow weeder, platform for stringing, equipment for the application of crop protection compounds, aircraft for applying insecticides,

5. harvest trailer for transport of hop bines, devices for cutting, separation and loading of hop bines, stationary hop harvester capable of dealing with 500–650 kg fresh hop per ha, subsequently, one able to handle 750–850 kg, high efficient dryer for 750, 1000 and 1500 kg fresh hop per hour, air conditioning equal to the efficiency of dryers, packing press.

TABLE 53   Traditional, contemporary and prospective management of hop gardens

Traditional method	Contemporary method	Future method
Autumn period		
All management methods	– clearing the garden – manuring with farmyard manure once every 3 years – spreading artificial fertilizers – liming once every 3 years – harrowing	
Autumn ploughing (ploughing-in or ploughing-away, i. e. from the stand of plants)	– ploughing-away or to be replaced by shallow loosening	– ploughing-out – shallow loosening instead of ploughing – deep cultivating once every 3-5 years
Spring period		
All systems of management	– repair and maintenance of structures – spreading artificial fertilizers – harrowing	
– ploughing-away – removal of soil – manual cut – rolling of inter-rows	– mechanized cut	– mechanical or chemical removal of prematurely sprouted shoots – application of herbicides
From the cut up to harvest-time		
All management methods	– continuous protection against diseases and pests	
– manual suspension of steel wires – first training of runners – ploughing of inter-rows – second training of runners – additional manuring with artificial fertilizers – first ploughing in – continuous training of displaced runners – cultivation of inter-rows 3 to 5 times – additional manure spreading – second ploughing in – continuous re-suspension of collapsed plants	– manual suspension of strings – training of runners – ploughing in – continuous training of displaced runners – cultivation of inter-rows 3 times – additional manure spreading – continuous suspension of collapsed plants	– mechanized suspension of strings – simplified training of runners – training of displaced runners – application of herbicides

Fig. 103. Traditional
cultivation schedule.

	Classical technology	
autumn work	clearing of bines spreading of artificial fertilizers extra ploughing or final ploughing once every 3-4 years ploughing -in dung or liming	
winter	repairs to structures	
spring work	harrowing ploughing-away ⎤ spreading removal of soil ⎬ of artificial from the row ⎭ fertilizers of plants finish and cutting, rolling	
treatment during the vegetative period	suspension of strings dusting to control hemp flea beetle first training leading to final shaping weeding second training first spraying against downy mildew first ploughing – in 2 to 3 sprayings weeding 2-3 times additional application of nitrates second ploughing- in spraying of cones	
	harvest	

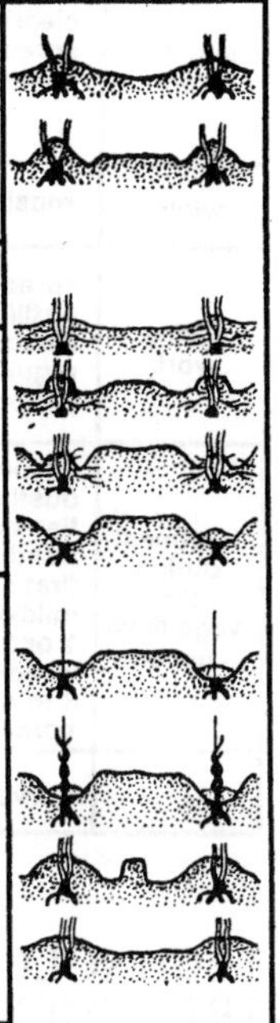

	Present practice	
autumn work	clearing the remnants of bines spreading of artificial fertilizers ploughing-away or loosening once every three years ploughing-in of dung or liming	
winter	repairs to structures	
spring work	spreading of artificial fertilizers harrowing mechanized cut	
treatment during the vegetative period	suspension of strings dusting to control hemp flea beetle training first spraying against downy mildew weeding first extra ploughing 1 to 2 weedings 3 to 4 sprayings additional application of nitrates second extra ploughing spraying of cones	
	harvest	

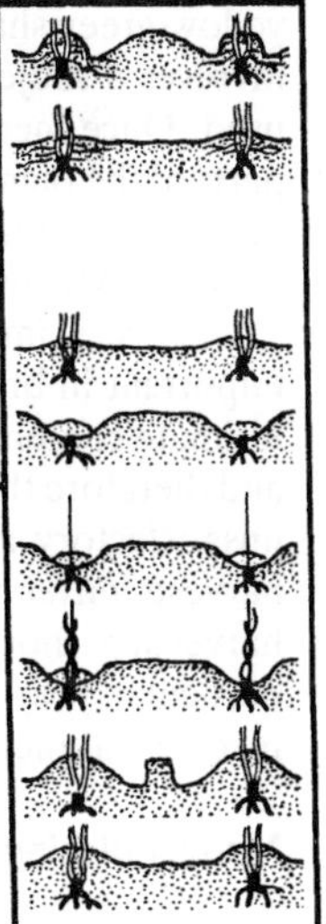

Fig. 104. Present-day
cultivation schedule.

221

	Flat area	
autumn work	clearing the remnants of bines spreading of artificial fertilizers loosening or ploughing	
winter	repairs to structures	
spring work	spreading of artificial fertilizers harrowing regulating the sprouting period of runners	
treatment during the vegetative period	suspension of strings dusting to control hemp flea beetle training first spraying against downy mildew 1 or 2 loosenings – weedings 3 to 4 sprayings, last spraying of cones	
	harvest	

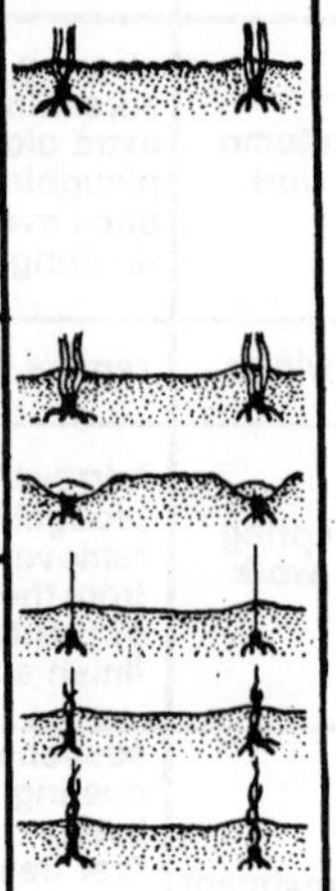

Fig. 105. Cultivation schedule in flat areas.

TECHNOLOGY OF HARVEST AND AFTER-HARVEST TREATMENT OF CONES

Hop picking

Technical maturity is usually based on the tightness of the hop cone, its elasticity, its typical yellow-green shading, its lupuline content and the typical fine hop scent. The timing of harvest involves a subjective evaluation which was sufficient in the past, when manual picking was used. Once mechanized picking was introduced, then the speed at which the above-ground parts of the plants dry out, must be taken into consideration. Hop plants are usually ready for mechanized decapitation a little later than for manual picking, because immature cones die back quickly and therefore the number of damaged and lost cones increases.

It is very important to know the dynamics of the development of those components important in brewing because their content and levels has recently become important. The ultimate contents of these components become fixed during the final few days before harvest and therefore the old method of evaluating the state of technical maturity has recently become unsatisfactory and will not serve in the future. According to HAUTKE (1976) individual harvests differ with regard to the dynamics of production of the bitter acids. Therefore, harvesting should start as soon as the content of bitter acids becomes stabilized in that particular year. Therefore, the methodology for the prognostic regulation of the start of mechanical picking is under investigation.

Manual picking

This method of picking is slowly coming to an end under present day conditions in Middle Europe. It involved the manual removal of the hop cones and it was required that the cone should be complete, including its pedicel or that one peduncle should carry a maximum of three cones. Picking was done within the hop garden into large woven baskets which were carried to the central weighing point. There the cones were tipped into the measuring vessel and at the same time their quality was judged. The measuring vessel a had a capacity of 30 l. In some hop-growing regions so-called 'big quarts' with a capacity of 60 l were used. After measuring the hops were pressed into big sacks, the so-called bales. An unexperienced picker would harvest approximately 10 quarts in a shift, an experienced picker, roughly 20 quarts and a top performer could reach 40 quarts.

Manually picked hops had a lower total density because they included pedicels, so they settled less and were suitable for short-term storage in their fresh condition (maximum 6 hours). This method is very inefficient and, therefore, is going out of use. The harvesting of one ton per ha at an average of 10 quarts per picker per day, over a 15 days harvest period, required 12–14 handpickers. In Czechoslovakia manual picking, in the past, was mainly done by middle school students and apprentices; the total number of seasonal workers required in the whole state was 100 to 130 thousand.

Mechanical picking

The disadvantages of manual picking, particularly the enormous labour requirement led to the introduction of mechanized picking. The first experiments with machine picking were mentioned in the early 1900s. California was the cradle of mechanized harvesting (MOHL, 1924). Different methods of mechanical harvesting appeared between the two World Wars, and English manufactures were the most successful, their products being exported to various countries after World War II.

Different principles for removing the cones from the hop plants, such as cutting, plucking and knocking, were investigated in Middle European conditions. The question arose as to whether the machine should be stationary or should be mobile and work directly in the hop garden. The cleaning of the picked cones underwent a complicated development before it was finally standardized.

In 1954 large hop-growing enterprises came into being and required the replacement of manual picking by mechanized harvesters because of the large areas to be managed. Therefore, English machines (manufactured by Bruff and Robotank) were imported and their effectiveness was investigated with Czechoslovak hops because these have specific requirements for mechanized picking because of the fine structure of the cone. Once the mechanization of hop picking on Czechoslovak conditions was verified a derived type of picking machine was developed and the first type (ČCH 1 = hop harvester 1), was tested in 1959. The mechanization of hop picking presented many problems which were reflected in the efficiency and quality of the picking machine. The development of new Czech types, accompanied by their testing, and comparison with machines available on the European market, proceeded at good speed. The following imported and Czechoslovak machines are currently avaible: Bruff-D-9 (England), Bruff D-11 (England), Bruff E (England), Bruff H (England), Week Simplex (England), Rotobank Baby (England), Hinds-Rotobank 2/40 (England), Allayes M 22 (Belgium), Allayes M 42 (Belgium), Wolf 700 (Federal Republic of Germany), ČCH 1 (Czechoslovakia), ČCH 2 (Czechoslovakia), ČCH 3 (Czechoslovakia), ČCH 4 (Czechoslovakia), LČCH 1 (Czechoslovakia).

Czechoslovak machines were successful and became the basic type for Czech hop growing. They have been exported to other Comecon countries, in which Czechoslovakia is now the sole producer of hop harvesting machines.

The development of mechanical harvesting in Czechoslovakia is shown in Fig. 106.

In 1976, 92 per cent of the total area was mechanically harvested. Therefore, it can be said that there is full mechanization of hop harvest, because only a small proportion of young hop gardens is not harvested by machine.

Its latest type of hop harvesting machine recently produced in Czechoslovakia is LČCH 1 (line – hop harvester No 1) and LČCH 2 is now in production.

The harvester type LČCH 1 is a stationary picking machine able to harvest 500–650 kg per hour or 20 ha in a normal season. It has a power consumption of 25 kW and an operating staff of 20 men. It consists of double automatic equipment with conveyer, cleaning and sorting equipment, controlling conveyers, containers suspenders, and insertion device. The picking system is based on the insertion of vertically suspended bines and unlike the Bruff. In this system the bines here are pulled upwards. This equipment, which consists of 4 picking drums each with 6 picking bars, is driven by two electric motors. The rotation of the upper and lower drums is independent and can be smoothly altered, while operating, by means of a belt controller. The picked bines are moved onwards by a chain conveyer to a cutter equipped with separating drums. The picked mass then passes onto a conveyor belt to a harvest finisher the function of which is to collect the larger clusters of cones picked earlier. The picked cones fall on to a moving belt and are transported to the cleaning equipment

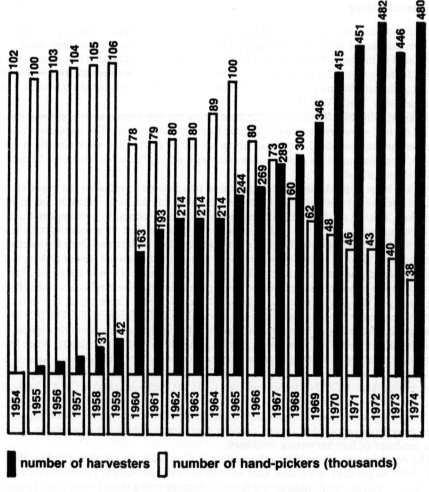

Fig. 106. Development of manual and mechanical picking in Czechoslovakia.

Fig. 107. Hop picking machine type ČCH-4.

TABLE 54 Performance of hop-picking machines

Type of machine	Number of machines working in Czechoslovakia			Output of machine operating during the whole harvest season (in tonnes of fresh hop)			Output of machine operating per hour (in kg fresh hop)			Number of hectares harvested by one machine		
	1975	1976	1977	1975	1976	1977	1975	1976	1977	1975	1976	1977
CCH1	21	18	14	64.00	43.37	63.16	250	229	185	14.21	12.14	9.78
CCH2	42	39	38	67.96	59.19	80.53	283	302	319	12.60	13.10	12.90
CCH3	133	137	141	73.06	63.41	95.03	271	268	302	15.98	15.73	15.09
CCH4	124	120	121	94.73	76.67	111.20	328	325	358	19.75	19.21	17.90
LCCH1	74	102	121	96.36	76.09	116.35	359	355	357	19.70	19.78	19.52
Bruff D-9	13	10	19	60.06	47.00	83.35	300	215	290	13.28	13.85	15.02
Bruff D-11	24	26	22	63.34	52.68	83.50	184	245	264	19.38	15.27	14.09
Bruff H, E	7	7	7	131.83	96.77	129.91	483	417	450	26.14	21.88	22.50
Allayes M 22, 42	37	37	38	79.17	65.08	92.73	298	278	275	17.62	17.94	15.37
Wolf 700	1	1	1	107.00	45.00	86.10	450	201	283	20.00	23.00	25.00
LCCH	–	–	1	–	–	171.62	–	–	488	–	–	25.43
Total average	486	500	525	83.80	62.53	100.59	321	284	322	17.87	17.19	16.46

Leaves and picked shoots are sent by conveyor to a second waste material collector where it passes, via a hopper to a cutter.

There are now more than 500 machines at work in Czechoslovakia. Their efficiency is shown in Table 54.

The mechanical harvester is not merely a piece of equipment, but it has an effect on other areas of hop technology. It is involved in the question of pre-harvest decapitation, it influences the quality of picked hops and is concerned in various aspects of crop losses.

Harvesting decapitation

Pre-harvest decapitation is a violent intervention which interrupts the life process of the hop plant and causes permanent effects. RYBÁČEK (1960) does not recommend this intervention for young plants in the first or second year after the foundation of the hop garden, because it is at this time that the permanent underground organs are formed.

Harvest decapitation attracted attention long before mechanized picking was introduced and was discussed when hops were grown in pole gardens, and where the bines were regularly cut for harvest. KODYM (1879) recommended the later cutting of young plants. VÁGNER (1910) noted a decreased yield by 23.9 per cent after pre-harvest decapitation, ZATTLER (1954) quoted reductions of 11 to 46 per cent and RYBÁČEK (1960) 4.32 to 35.57 (where the reduction was proportionally less with increasing height of decapitation), and HAUTKE (1960) 6 per cent (with a possible maximum of 30 per cent).

Despite this disadvantage of harvest decapitation, there is nothing to replace the mechanical harvester. Therefore it is important to determine how and when this operation should be performed so as to minimize crop losses. The following aspects need to be considered:

1. Young stands should not be mechanically harvested in the lst and 2nd year after the creation of the hop stand from cuttings.

2. Delay the start of the harvest, because the later the decapitation the smaller will be the yield losses in the following year.

3. Organize the harvest so that the stands picked first in one year will be picked last in the next year and make a similar arrangement within individual hop gardens.

4. Reduce the harm done by decapitation by leaving at least part of the assimilating area on the plant, i.e. cut, if possible, just below the lowest shoot bearing cones.

5. Regulate plant growth and aid the production of hop cones by the planned application of nutrients.

The quality of harvested hops is associated with the problem of losses. Thus involves different degrees of damage to the cones and means the production of the highest possible number of cones apparently undamaged as seen by the naked eye.

The testing of picking machines by the state distinguishes several degrees of damage:

First category – undamaged, cones, with the visible lower part of the spindle intact together with a remnant of pedicel with bracts growing from it. It is assumed that this quality will be preserved during the subsequent, after-harvest, treatment. This category should be the most numerous.

Second category – cones slightly damaged. The tips of the bracts may be damaged.

Third category – parts of the cones totally decomposed. This category is the most unwanted, as during the harvest and in the post-harvest treatment the losses increase because the lupuline content decreases and in the brewery the cones disintegrate, making filtration difficult and fouling the filters.

Judgement of the quality of the work of these machines must assess the quantitative and qualitative indicators of their function, including the various preparatory operations.

Losses of quantity and quality can occur at various times in the harvesting process:

1. during decapitation, pulling down the plants, and loading and transporting them to the harvesting machines;
2. during the picking process itself;
3. while the picked cones are awaiting their next treatment.

As was mentioned earlier, the decapitation can affect losses, because the bine wilts before it gets to the machine if it is cut prematurely. Therefore a basic requirement is to cut the bine immediately before it is loaded onto the transport. When the plants are pulled off the strings, it is necessary to take the rests of plants from the wirework down and at the same time to clean-up around the loaded transporter. This can be done without difficulty by the day shift workers but is less easy at night. Therefore, it must be done as soon as possible after the start of the day shift. Losses from the transports between the hop garden and the harvesting machine are usually small. Often the plants are carelessly loaded on the transporter so the cones become dirty and have to be discarded when suspended on the harvesting machine. Losses within the hop garden and during transfer to the picking machines should not exceed 1 per cent of the total weight.

Quantitative and qualitative losses can also occur at the picking machine. The functioning of the machine depends to some extent on climatic conditions and on day length. It also depends on the shortest possible time elapsing between the cutting the bine in the hop garden and its insertion into the machine. The best conditions are those of low temperature and high relative atmospheric humidity, which within a 24 hour cycle, usually occur during the night; the worst conditions are usually at noon and during the early afternoon.

The operation of harvesting machines is also affected by the growth habit of the plant. A close ratio between the mass of cones and the other biomass of the plant is the ideal condition for mechanical harvesting. Different varieties differ in their suitability for machine picking. A very important property is good resistance of cones to mechanical damage.

The amount of loss at the machine can be reduced by careful insertion of plants into the picking section. The perfect function of this section is vital in maintaining the high quality of hops and reducing losses. of concern here are the basic principles of the picking equipment (horizontal insertion, vertical insertion, with subsequent movement of the plants upwards or downwards), together with its ease of maintenance and adjustment during operation. The requirements of its operation are always precise, because any damage caused to cones during this phase cannot be corrected during subsequent phases.

The succeeding section in Czechoslovak harvestors (ČCH and similar machines) involves the separation of the various components after picking; the picked cones have to be separated from them or their remnants so that these unwanted components do not arrive, in great quantity, at the automatic harvest-finishing section. There the cones may suffer severe damage, if they reach the harvest-finishing section already damaged during passage through the picking section.

The next phase is where the lighter plant debris is removed. This is the so-called foliar waste material, which has to be cleanly separated from the cones. Here, different principles tend to be used (e.g. pneumatic or mechanical) but their function is simply to clean the hop cones and to remove the unwanted components without any damage to the cones. Even here the rate of loss is considerable, mainly due to a lack of reliability in separation.

The total losses in mechanical harvesting should not exceed 7 per cent. In fact total losses are usually greater than this, reaching approximately 10 per cent and where there are malfunctions of the machinery losses be as much as 30 per cent.

The organisation of the harvest has involved the introduction of shift work, so that the picking machines can be used in two-shift or three-shift operations.

A three-shift operation makes for better usage the equipment but it increases the amount of labour required. This maximum usage of technical equipment with its higher output makes it essential to discard old fashioned machines and to use only a modern installation.

Hop drying

Drying is the simplest method of hop conservation. It inhibits the respiration of the cone and reduces its moisture content to a level at which potentially damaging microorganisms are inactivated. Drying is as old hop-growing itself. Hop cones were dried in the distant past under natural conditions in thin layers in the shade. This method was very slow and was gradually improved by drying on different flooring materials (earth, bricks, wood). Later, hops were dried on lattice-work hurdles which could be positioned one above the other. The drying process was then accelerated when the rooms containing the hurdles were heated. This period marked a turning point, from the natural method to the thermopneumatic. The use of heat plus a forced draught made an orderly arrangement of hurdles a necessity and so the first hurdle-drying buildings came to existence (in Czechoslovakia e.g. Baŕtipán's drying building in Hŕedle, district Rakovník, Central Bohemia). The hurdles were stacked on the floor so that floor hot air passed through the lower layers to the upper. From this arrangement only a small step was necessary towards the driers used today.

Chamber driers

The original chamber driers were simple adaptations of malt kilns, used for drying malt. They had a perforated floor made up of louvres situated immediately above the source of heat. The fresh hops were laid in a thin layer on these louvres and the warm air passing through created the drying process. The hop layer was regularly raked so that whole drying process lasted only one day. These driers appeared at the end of the 1800s and have been continuously improved in some countries, e.g. in England, where they are used today with improved versions of the combustion chamber and vaporizer.

Further development led to the structure which served as the pattern for modern Czechoslovak chamber driers. Fixed louvres were changed for adjustable ones and small trucks with perforated bottoms (currently made of wire mesh) were placed under the louvres. It was thus made possible to dry the hops in two layers and to discharge the drier by means of the adjustable louvres. Different constructions appeared during this period, e.g. the louvres were replaced by trucks and the drying process was again done in one layer, or the trucks were filled with fresh hops which were regularly raked during the drying process. There were also modifications of the louvres and the number of hinged louvres increased (at first there was one hinged louvre to each truck and later there were two louvres per truck) with individual waste air ducts. The driers were usually set up in corn lofts and the drying space was in their top floor, under the roof, within the exhaust exit at the highest point of roof.

The combustion chamber was very quickly changed. At the start of warm air drying, the system involved either the hot products of combustion or simple heat exchangers. Where combustion products were used directly there were from charcoal combustion products were used directly there were from charcoal and coke, but notwithstanding certain advantages, especially, low cost, this method did not find acceptance in Middle Europe, where the simple heat exchangers were preferred.

Many types of hop driers were developed in the period between the two World Wars, but two principles attracted particular attention.

The first of these was featured the "Saazia" drier, where the hot air was led to the trucks and then to the louvres above them or was led through a side channel directly to the louvres. This method equilibrated the temperature loss of the air which had passed through the layer of hops in the trucks. The system has been gradually improved and is used at a better technical level today.

The second modification concerned the exhaust of the air from the drier. Early systems used forced circulation or even forced withdrawal of the air from the room above the louvres, where simple inefficient suction fans were usually installed (e.g. Trade Mark Lupulus). This was in fact, a very interesting and progressive construction.

After World War II there was a variety of driers available from 4 m² up to 30 m² and more depending on the size of the hop garden.

Before World War II, attempts had been made to build continuous driers, as distinct from the normal chamber driers (in those days called louvre-type or hurdle-type driers).

The change towards a collective economy in Czechoslovakia with an associated increase in the size of hop gardens created a need for driers of greater efficiency.

The first thought was to increase the number of louvres. However, this increase was limited, because such a system depends on a certain level of dryness of the cones being reached on the topmost louvre. If this requirement is not satisfied then moist hops are tipped onto the lower louvre and fresh hops are poured onto the topmost louvre. These hops will then receive moist air from the lower floors of the driers and would therefore, become heated while stacked, and would therefore loose their value. The maximum number of louvres shown to be usable was 3 plus 1, i.e. 3 louvres plus trucks.

The second possibility was to increase the floor space of the drier. The limitations of this solution soon appeared when it was found that an existing drier with a floor area of 30 m² had problems with the distribution of warm air. These difficulties led to the conclusion that a solution lay in the use of multi-chamber driers in which several (usually 3–4 but sometimes as many as 8) drying units were joined together.

The basic problems were mostly concerned with the size of the chambers and their air supply and air movement. The solutions were based on standardization to provide a chamber ground (5 × 4.5 m, i.e. 22.5 m²) trunking to supplying the drier with the air (0.16 m² per 1 m² drying surface), a minimum floor area of the vaporizer (0.6 m² per 1 m² drying surface), a minimum height between the lowest louvre and the combustion chamber (6 m) and a number of chambers (up to a maximum of 8).

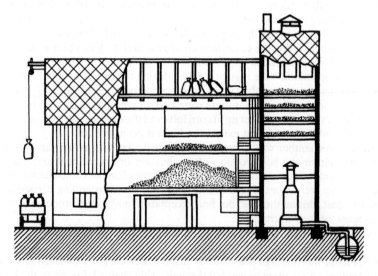

Fig. 108. Diagrammatic representation of the layout of a chamber drier.

Fig. 109. Discharging the hurdle from a chamber drier.

In the subsequent development the chamber driers were improved as follows:

1. A conveyer system was introduced for the transport of hops to the drier; this replaced the existing transport of bales which were taken up by a lifting apparatus.

2. The louvres were modified so as to increase their flexibility, and the trucks were replaced by continuous belts on which the drying process could be completed and which facilitated the discharge of the chambers.

3. The coal combustion chamber was replaced by a thermopneumatic unit with a burner for light mineral oil and with a heat exchanger in which waste hot gas was mixed with air supplied by axial fan.

4. Air was taken out of the drier by suction fans with a throughput 10 to 15 per cent greater than the amount of air supplied to the drier. This forced circulation of the air removed the dependence of driers on the weather.

5. These various modifications increased the output by 180–200 per cent. Therefore, it became necessary to introduce certain other post harvest treatment of the cones. The storehouses were equipped with a conditioning chamber connected to the drying units (the principle and description is found on page 242).

Belt driers

As well as the *louvre-type driers,* other types have appeared. The most interesting among them were those constructions intended for use as a *continuously-operating drier* (as e.g. the tunnel drier erected on the farm in Tuchořice and produced by the Blaschke company, with moving hurdles and forced air circulation, or the eleven-band drier on Pihrt's farm at Pnětluky, district Louny). These designs were not adapted elsewhere because their construction was not perfect and they were not suitable for the small-scale production common at that time.

Fig. 110. Modern four chamber drier with the hop conveyer above the chambers.

Fig. 111. Modern four-chamber drier with manually operated louvres.

The need for a continuous drier became obvious when machine picking was introduced because this method produces damaged hops which cannot be stored for very long in the fresh condition. So, in 1959, the first band drier, built by Binder and Bürgmayer, was imported into Czechoslovakia and erected on the state farm at Žatec, at the Břežany plant. It was a three-band drier with vertical combustion chamber and its burner could utilize fuel oil or coke. When the continuous drier was tested under operating conditions on a large scale it was shown that this type of drier is suitable for Czechoslovak hops.

Therefore a factory in Nové Mesto nad Váhom was modified for the production of band driers. The early models, code numbered PSCH-325, had a combustion chamber for using coal (1961), but this soon changed in favour of combustion chambers for light fuel oil. The demand for such drying equipment was greater than the production capacity and therefore driers (Vojvodianka SH 200) were imported into Czechoslovakia from Yugoslavia.

The production capacity of the band driers manufactured in Czechoslovakia is rated so that

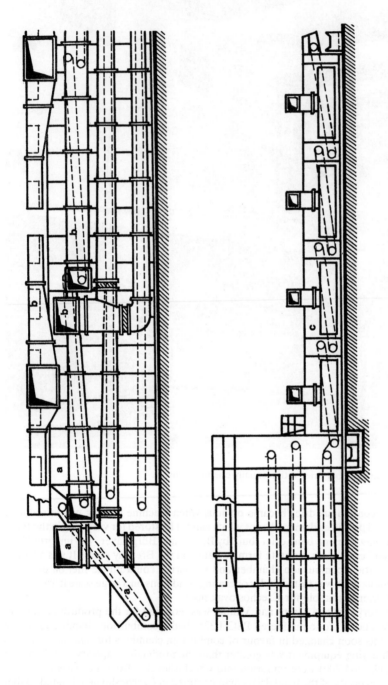

Fig. 112. Diagrammatic layout of a large-scale band drier: a – input line, b – band drier, c – air conditioner.

Fig. 113. Band drier PCHB-750 with air conditioning chamber (in foreground).

it corresponds to the capacity of picking machines and thus makes it possible to establish complex picking centres. First types of band driers had the capacity of 325 and 375 kg of fresh hop per hour. The growth of the capacity of picking machines led to the design and production of driers with 500 and 750 kg per hour capacity. In present times driers with a capacity of 900 and exceptionally even 1500 kg fresh hop per hour are available.

The main part of any drier is a drying box fed with warm air from an air warmer. In this warmer, the pollution of air by used warming media must be avoided. The warming box is provided with three band conveyors mounted on three floors. The lowest band is supplied with the main stream of drying air. The stream of drying air fed to the other two bands is less voluminous. The speed of these three bands is not the same. The bands in upper floors move more rapidly than the bands on lower floors. Because all bands are supplied with the same quantity of hop, the upper bands carry thinner layers of fresh hop and the lower bands layers of successively dried hop. The final drying procedure takes place on the lowest band.

The air which passed through the layer of dried hop is taken out from the drying box by a suction fan, whose piping is mounted in the top part of the drier. Because recycling of drying air is not suitable, the whole quantity of the air is taken out. Part of this air is used for an air conditioning chamber, which is usually joint with the drier.

The technology of hop drying

Three phases of the drying process can be distinguished:

a) heating the material: here the supplied heat raises the temperature of the hops and this phase lasts until the supplied heat is in equilibrium with the heat of evaporation;

b) constant speed of drying: in this phase the temperature of dried material is constant at the same level as it was at the end of the heating phase. This phase lasts to the time of so-called critical humidity when the surface of the dried material behaves like a free water surface and the vapour tension above this surface is equal to the vapour tension above the free water surface;

c) reduced drying speed is a phase in which the supply of water from inner parts of the dried cones to the surface is insufficient when the critical humidity is reached; it ends when the contained moisture reaches equilibrium.

The drying process is affected by many factors and lasts 5–8 h or more. It has its own peculiarities and is regarded as the most important operation in the harvesting process.

Four basic parameters which affect the drying procedure and its result are the following:

1. specific drying properties of the hops being dried;
2. drying temperature;
3. volume of air and the speed of its movement;
4. other factors.

Specific properties of hop cones in relation to drying process

In terms of the drying process picked hop cones can be regarded as a living organism whose basic life processes, particularly respiration, are continuing. They first react to being removed from the plant by a higher intensity of respiration. This feature is further intensified by injuries due to picking, which can be numerous if mechanical picking is used. Here two different levels of injury can be distinguished:

a) *macroscopic injury*, clearly evident as damage to the whole structure of the cone such as broken cones and missing bracts and eventually cones are seen to be decomposed;

b) *microscopic injury*, not evident to the naked eye, involves fine injuries to the surface of cones due to contact with the metal parts of the picking machine or due the mutual friction of different parts of hop plants during the picking and harvesting process.

If both degrees of mechanical injury are considered then there can be hardly any cones without injury. Most importance, however, is given to macroscopic injury, which has evident consequences during the subsequent treatment of the cones. From this viewpoint the cones can be classified into different categories based on the quality of the product of the picking machine.

Only those cones which have at least a part of the pedicel or have an undamaged basal segment, and which are tightly closed, with regular bracts covering each other (more of them lost or evidently damaged) are classified as undamaged. The volume of this category depends essentially on the technical standard of the picking and harvesting equipment, on the hop variety, on the degree of maturity of the cones, on the particular year and locality as well as on the level of husbandry. This highest quality of relatively undamaged cones ranges from 50 to 80 per cent of the total weight.

The proportion of damaged cones is thus fairly high and this has its consequences before and after the drying process. The increased respiration associated with damage frees moisture and releases energy so that the temperature of cones increases as well as the humidity at their surface. This causes fustiness.

Injuries which occurred earlier in the hop garden, have to be considered. These include

236

damage caused by hop aphid and spider mite, and the changes in colour and the reduced mechanical resistance caused by downy mildew.

Hop cones carry their specific microflora which may so develop after harvest that they affect the quality indicators (brightness and overall colour).

The structure of hop cone is based on certain important features. Of particular significance is the spindle, which forms the main axis and is a mechanical stiffener, and also the bracts which cover each other and produce a barrier to the drying medium. The sizes of hop cones vary considerably from year to year as well as from place to place. The average cone length ranges from 25 to 30 mm but can be as little as 10 mm or up to 80 mm. Middle European hops have a relatively stable mean size. The mass of picked hops includes other components which need to be separated. These include petioles, bine and shoot leaves (whole or fragmented), pieces of broken branch, and finally different mechanical rubbish such as hooks, bits of wire, small stones, clods of earth and other totally foreign objects.

The chemical composition of cones is very important and must be respected during the drying process as well as in subsequent treatments. The essential substances are very sensitive to their environment because they volatilize or rapidly become oxidized. The importance of those substances essential for brewing was treated elsewhere in this work.

The technology of post-harvest treatment depends on other characteristics of hop cones, the most important of which are:

moisture content of picked hop . . . . 76 to 82 per cent.
density of fresh hop . . . . 66 to 87 kg per m^3
   a) manually picked hop . . . . 73.3 kg per m^3 is the most common value,
   b) machine picked hop . . . . 83.3 kg per m^3 is the most common value,
weight of 100 cones a) fresh . . . . 45 to 75 g,
              b) dried . . . . 10 to 17 g.

Drying temperature

Since warm-air drying was introduced careful attention has been given to the drying process. As a rule, this is the only process indicator discussed in older publications. The drying temperature usually refers to the temperature of the drying medium, which is usually warm air and only rarely is a mixture of waste gases and air. Quoted drying temperatures range from 40 to 60 °C , and sometimes higher and it seems that the differences lie in the quality of the equipment being used.

Presented here is a selection of drying temperatures according to particular authors: TOMEŠ (1891), 30 °C, TOMEŠ (1924), 38–40 °C, ZÁZVORKA–ZÍMA (1938), 40 °C, maximum 45 °C, ZÁZVORKA–ZÍMA (1924), 55 °C, SCHUSTER–KREININGER (1955), maximum 60 °C, KUNZ–SKLÁDAL (1958), 40–50 °C, DREXLER (1961), 50–53 °C, BAILEY (1958), 65 °C, BAILEY (1963), 55–60 °C, maximum 65 °C, FRIC–MAKOVEC (1965), 55–60 °C. All of these authors agree that there is a maximum drying temperature limit, which, if exceeded, is likely to reduce hop quality.

In studying the reason for the d i f f e r e n t d r y i n g t e m p e r a t u r e s , it is necessary to realize that a general designation for the drying temperature cannot be given since this parameter must relate not only to the temperature of the drying air, but also to the temperature of the bulk of hops being dried and to the temperature of the individual hop cone, i.e. its internal temperature.

The temperature of the air is the most frequently quoted indicator, because it is the most easily measured. However this temperature is not always solely decisive as the final temperature effect on the quality of the dried hop. The most important question is when, and under what conditions, the temperature of the drying air is equal to the internal temperature of the hop cone. A high air temperature need not always mean that the temperature of the heated material reaches the same level. Of vital importance here is the time factor, i.e. the period of time over which the hops are being subjected to hop-air drying, i.e. the interval before the internal temperature of the cone reaches that of the drying air. This condition of thermal equilibrium itself also depends on many factors, including the speed of flow of the air, the humidity of the drying air, the moisture content of the hops, and the penetrability of the hop layer.

If the data given by the various authors are seen in this context, then it is easy to arrive at an explanation for their differences. Obviously, the most important factor affecting these ratings was the speed of the air flow. Older style driers with no forced ventilation had a low air flow speed; in this case, the ratings mentioned would have a lower limit. The drying air temperature, which is still the most often quoted indicator, can, therefore, be used to describe the drying process, but only related to the out-flow conditions of the equipment being discussed.

In recent literature temperature readings are given which may be designated as the temperature of dried material mass. This is usually determined by enclosing the dried hops in a thermally insulated vessel and measuring its temperature with an ordinary thermometer. It will be apparent that the temperature of the material, as compared to the temperature of the air being fed into the vessel, will be lower because heat will be absorbed by the dried hops cones. This difference depends primarily on the moisture content of the cones during the time of measurement.

The internal temperature of cones is the most difficult to determine because it is dynamic. Precise data are not yet available and the respective relationship is under investigation.

The air temperature suitable for hop drying must, therefore, be determined in relation to the speed of air flow. If, in a stable situation the air flow speed is 0.5 m.s^{-1} drying equipment is designed to move the hops at 0.45 m.s^{-1}. At this speed the drying air temperature can reach the maximum level of 60 °C without any ill-effect on the quality of the hops. However, this temperature level is governed by the specific qualities of the cones, primarily by their level of maturity. Therefore, it is recommended that a lower drying temperature (55 °c) is used at the start of harvest and to increase it to 60 °C as the more mature hops are picked up.

The question then arises as to whether it is possible to raise the temperature of the air with a higher speed of flow. According to the Binder Company, with a flow speed of over 1 m.s.$^{-1}$, it is possible during the early drying period to use temperatures higher than 80 °C. But the experiments of FRIC and MAKOVEC (1976), repeated over several harvests, did not confirm this assertion although their investigation involved a graduating flow speed.

The effect of temperature on hop quality can be evaluated by eye according to *the lupuline colour*. Provided that the temperature of the heated material has not exceeded the maximum, the lupuline colour remains lemon yellow. The ill-effect of a high temperature will first become evident as this colour changes to brown. The higher the temperature and the longer the exposure period the more apparent is the brown colour of the lupuline. This indicator is relatively reliable and is so simple that its value as a check on the drying process is quite obvious. This indicator has a direct relationship with the chemical content of the hop. Chemical analysis will demonstrate the negative influence of a high temperature by showing a higher content of hard resins. The proportion of these substances should be as low as possible and must not exceed a maximum of 2 per cent immediately after harvest. Before there is a change in the chemical composition of the cones, the change of lupuline colour is very apparent. The glossiness of cones is also sometimes quoted as a criterion of the applied temperature. But changes in this quality is not necessarily an indicator of the drying temperature, because it can occur when fresh hops are stored before drying without mashing or when dried with an insufficient air supply. Though changes in the overall appearance can be caused by a high temperature, it is always necessary to consider all possible causes and effects.

238

The volume of air and its flow rate

The drying of hop cones is very demanding of the volume of air and on its speed of flow. The moving air has to transfer to the cones the heat required for water evaporation and simultaneously provide a rapid exhaust of tne released moisture. Should this exhaust process be insufficiently swift and effective, the relative humidity would increase within the layer of hops with a consequent loss of the gloss and a change in the colour of the cones, first to yellow and eventually to dirty yellow.

Because with most drying equipment used for hops the speed of air movement cannot be regulated the colour must be considered as the limiting factor so that the temperature and consequently the level of water evaporation from the cones has to be adjusted accordingly. Thus, the basic principle of current hop drying equipment must be criticized from this standpoint. At present only *counter-current driers* are employed (the layer of hops moves in the opposite direction to that of the drying air), whereas a *co-current system,* though as yet unpracticable, would be of great advantage with regard to the efficient use of air. The counter-current system is also the reason for the relatively low saturation with moisture of the air discharged from the dryer, but the air cannot be recycled. There would be nothing gained by mixing even a part of the air discharged from the drier with that to be fed below the layer of drying hops. A relative humidity limit has not yet been quoted for the warm air supplied into the dried hops. It is generally assumed that the lower its relative humidity, the better is the resultant quality of dried hops and the higher the technological reliability of the drying equipment.

In modern driers about ± 140 m^3 of air are used per 1 kg dried hop. *The flow rate within the drier, i.e. the penetration through the layer of drying hops, varies according to the drier type as follows:*

in chamber driers (older type, i.e. the so-called farmer's driers) from 0.1 to 0.2 m.s^{-1};

in chamber driers (standard size) with natural circulation, from 0.15 to 0.3 m.s^{-1};

in belt driers (older type) with forced air circulation, from 0.3 to 0.4 m.s^{-1};

in present day belt driers and in modern chamber driers 0.45 m.s^{-1}.

In the design of a drier it is essential to fit the capacity of hops to the quantity of supplied and discharged air. Thus, to ensure the complete extraction of moisture from the layer of hops it is necessary to install an extractor capable of moving about 10–15 % more air than is taken into the drier.

It is usual to use warm air in hops drying, i.e. to install heat exchangers of waste gas and air. In spite of the fact that hop cones are very prone to absorb foreign odours, sometimes a mixture of waste gases and air is used for drying. In the past this method was used with coke heating. Today a mixture of air with natural gas or city gas waste is sometimes used. In all such cases, however, overall efficiency depends the thoroughness of combustion.

Other factors associated with hop drying

Other technological factors involved in hop drying include thickness of the layer of hops, the drying time interval and the speed of passage of the hops through different segments of the drier. Generally it is possible to determine such factors from the parameters already discussed, from which they result or according to which they have to be adjusted. Here, it should be quoted that there will be differences between individual years and so it is not possible to define standard reproductible conditions.

Post-harvest treatment of dry hop

Properties of dried hop

As it relates to the drying equipment and in terms of postharvest treatment, the hop cone is a very complicated structure. The drying process is not homogeneous and three separate phases can be distinguished:
1. dryness of bracts;
2. dryness of pedicels;
3. dryness of spindle.

The drying of the bracts does not proceed linearly and depends mainly on the level of contact with the streaming air. Because they have a large surface area and are thin they mostly dry out in the first phase although their bases, attached to the spindle are in contact with moist material which does not dry until the third phase of the process.

Where *the pedicel* remains as a part of the hop cone it becomes dried in the second phase. The degree of its dryness equals that of the whole cone.

The spindle is less accessible to the drying air than is the rest of the hop cone and therefore it is the last to reach the dried condition. Its drying proceeds from the base towards the apex and is regarded as dry when it becomes fragile.

The most reliable indicator of the level of dryness is moisture content. However, in working conditions judgement by eye of the condition of the cones is unavoidable. Different levels of dryness used for hop cones are compared with their actual moisture content in Table 55.

TABLE 55 **Classification of state of dryness of hops as related to actual moisture content**

Characteristics of hop cones	Moisture content of cones (per cent)	Technical term used for the state of dryness
The whole pedicel breaks, the lower segments of the spindle on medium-sized cones loose their elasticity and break. Moisture content of the sample varies considerably according to their homogeneity as to size. If there is a large proportion of bigger cones then the moisture content may exceed the limit of 11 per cent. The moisture content of equal-sized cones is usually 8 to 10 per cent.	8–11	including pedicel
The whole spindle is dry and liable to break. The apical segments of the spindle are elastic only in exceptional cases of overgrown cones. The bracts can be easily separated from the spindle.	5–7	including spindle

Recently, hops have begun to receive further treatment immediately after drying and therefore must be dried down to 5–7 per cent moisture. If the hops are dried to below 5 per cent moisture they are over-dried and this has a harmful effect on their chemical composition.

Dried hop is brittle and has low resistance to damage during handling. The bracts tend to fall from the spindle with a resultant loss of lupuline glands. The tendency of hop cones to disintegrate also depends on the temperature. If the temperature falls below the freezing point the cones are liable to disintegration even when their moisture content is above 10 per cent. This increasing tendency of hop cones to disintegrate as the temperature falls was proved in experiments by FRIC (1965) which assessed the effect of temperature on the mechanical resistance of hop cones down to −20 °C.

Disintegrated cones cause severe problems in breweries because they choke the filters. Neither are they suitable for processing as ground hops because the presence of a large proportion of stalks reduces the quality of the product.

Dried hop cones are hygroscopic and rapidly take up moisture from their environment. This feature is used in the post-harvest treatment for so-called climatization (see page 238). They are also photosensitive and, therefore, long exposure to light changes their biochemical structure as is shown by a typical red-brown colour, which is commercially undesirable. Hop cones are also prone to take up and absorb the scents of their environment and the presence of ammonia in the air spoils them by causing an undesirable colour change to a yellowish shade.

Treatment of dried hop

Dried hop cones require immediate treatment. If the final moisture content, after drying, is high, as was usual under small-scale production conditions, then the consignment should be ventilated and turned several times to ensure an even distribution of moisture throughout the whole bulk loaded mass. Moisture is not homogeneously distributed within the individual hop cones and they are considerably hydroscopic, therefore since microorganisms are not likely to have been exterminated in the drying process there is at this time considerable danger of microbial activity. This would endanger the commercial quality, so only hops dried to the stage known as "to the pedicel" can be put into short-term storage, and only then if the storage facilities are large and airy.

If hops are dried "to the spindle", then they are inclined to disintegrate. In this case it is possible to make use of its hydroscopicity and to apply such treatments as will ensure sufficient contact with air of higher relative humidity. In the past, natural conditions were used in the treatment of hops. Immediately after drying they were spread in small heaps of 50–60 cm high in granaries that were very well ventilated particularly during times of high atmospheric humidity. After 24 hours these hops were collected in a larger pile, usually 1 m high; this was continuously shovelled up and under favourable conditions it was possible to make the heap 2 to 2.5 m high for storage or to compress the mass into bales.

This method of treatment was dependent on climatic conditions. The frequency of years with low relative atmospheric humidity during the hop harvesting period is relatively high in Czechoslovak hop-growing regions and so the moisture content of the hops was rarely such as to allow the crop to be handled and compressed even with the best care and attention. Therefore it was necessary to increase the moisture content by sprinkling with water. But this intervention always involves certain risk, because the amount of water has to be exactly determined, the sprinkling very carefully performed and the hops have to be given further treatment by shovelling. Since the moisture content of cones is dependent either on climatic conditions or on sprinkling with water such a means of processing is not suitable for large-scale production.

Therefore the Czechoslovak institutions elaborated a processing method plus the necessary equipment for climatization or air-conditioning which adjusts the moisture content of hop cones in such a manner that they do not tend to disintegration. First, the moisture content of the cones was correlated with the humidity of their environment and a curve of equilibrated humidity was produced (see Fig. 114).

Then the speed with which the mass of dried hop cones absorbs moisture was determined and correlated with temperature before attempts were made to find a technical solution.

This study was based on the fact, that hop cones can be compressed and put into bales when their moisture content is less than 10.5 per cent. As Fig. 114 shows the required moisture level can be reached with relative atmospheric humidity of 65 per cent. All parts of the mass of hop cones must have access to this level of atmospheric humidity and this is a necessary condition, which must be respected by the equipment. Therefore, it was necessary to use forced circulation of the air and to arrange such a thickness of the layer, that each cone, which has its own uneven absorption characteristics would reach a state of balanced moisture content.

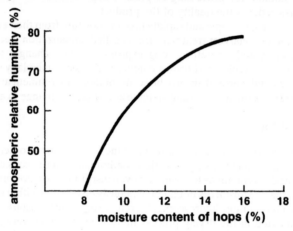

Fig. 114. Graph to show the relationship between the moisture content of hops and the relative humidity of the atmosphere.

To reach the required level of atmospheric humidity was a difficult technical problem. It was not possible simply to use the natural atmospheric humidity, because this varies during the daily cycle as well as in different years. Therefore it was necessary either to design equipment for the independent control of relative atmospheric humidity or to find another solution to the problem. A particular advantage was offered by the waste air from the hop drier, because this, particularly in the case of band driers, has almost stable temperature (38 to 40 °C) and relative humidity (roughly 45 per cent).

After a few years of experiments and pilot plant operation the treatment was finally arranged by passing the waste air from band driers over a water scrubber where it is cooled to 25 °C and its relative humidity is adjusted.

The equipment consists of an acclimatization chamber with air conditioning equipment. The acclimatization chamber is a tunnel the length of which corresponds to the required output. Within the tunnel are two conveyer belts; the first connects with the lower band of the band drier or with a collecting hopper in the case of a chamber drier. About half way along the tunnel the hops pass on to the other conveyor and during their passage through the tunnel the moisture level of the hops is regulated to the required level. At the end of the acclimatization chamber the hop falls onto a pocketed conveyor and is transported to the press.

The equipment for adjusting the relative humidity of the air consists of a collecting pipeline leading air, via a suction fan, from the band drier to the water scrubber where it passes through a water spray. This sprayer can be mechanically or automatically adjusted to provide the required relative atmospheric humidity. The water is collected and is recycled. The treated air is led by pipeline to below the moving belt, it passes throught the layer of hop and then exhausts to the exterior. In acclimatization equipment linked to a chamber drier, the waste air

242

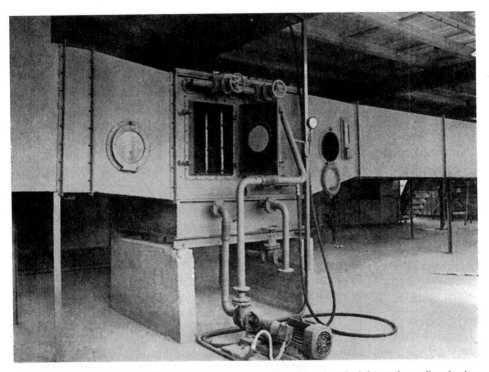

Fig. 115. Water scrubber for adjusting the relative humidity of the air at its inlet to the acclimatization chamber.

coming from an individual chamber, is combined before being sent via an independent fan to the water scrubber.

Only an approximation to the required moisture content can be achieved under operational conditions. The moisture equilibrium is not arbitrary for the final treatment of hop. The minimum moisture content, which has to be obtained during the passage of hop through the conditioning chamber, moves between 10.5 and 11 %. It has been found that the rate of absorption increases for a time, but then it decreases and equilibrium is reached very slowly. If the condition of equilibrium is to be reached, the time interval should be longer, approximately doubled, and the dimension of the equipment should be proportionally larger too. It has been found in experiments that it is neither necessary nor required to achieve complete equilibrium. This is borne out by the dynamics of uptake of moisture by the hop cone. The first part which absorbs the moisture is the bract and the last the spindle. It is however important that mechanical resistance to the disintegration is low, when the bracts begin to take up moisture, i.e. before there is an increase in the moisture content of the spindle. This explains why such a treatment can be regarded as a stabilizing factor, the quality of acclimatized hops. If the humidification treatment ceases at this time, the moisture will later disperse within the cone, i.e. the spindle will take moisture from the bracts. Theoretically therefore the quality of hop cones will not be reduced by a high moisture level, if it does not exceed a certain defined limit.

Main technological principles involved in the conditioning of hop cones

Technological problems involved in adjusting the moisture content of hop cones can be summarized as follows:
1. achieving complete dryness of the cones;
2. maintenance of the process;
3. correct treatment of acclimatized hops.

Complete dryness of hop cones implies a moisture content of 5 to 7 per cent, and this is a basic condition. It is the spindle which has to be completely dried out, otherwise the bracts would take water from the spindle during the period in the acclimatization chamber. Furthermore the spindles in the largest cones demand particular attention because they could be a source of moisture for medium and small cones. If drying is unsatisfactory then the baled hops would deteriorate by losing their sparkle (gloss) and their original typical colour.

Maintenance of the process means particularly the regulation of the relative humidity of the air. Relative humidity is measured by means of psychrometers in the inlet piping leading to the acclimatization chamber. An increased moisture content of the hops can be achieved by different methods, but a large and sudden increase of relative atmospheric humidity is not permissible, because it always leads to reduced quality. Under operating conditions the relative humidity of the air fed to the acclimatization chamber should be between 65 and 70 per cent and the time for acclimatization between 70 and 90 minutes. Under these conditions assuming that complete dryness had been reached, the moisture content of the cones will not exceed the determined limit, and will range from 10.5 to 11 per cent.

Hops moistened by passage through acclimatization chamber *must be immediately pressed into bales*. They must never be stored for long in bulk, waiting to be pressed into bales, because their moisture level can fall or rise according to the humidity of the environment. It should also be borne in mind that humidity measured by contact hygrometers will, in the main, be that of the bracts and this can be seen as apparent fall in the moisture level of 0.5 to 1.0 per cent. In fact the humidity remains on the same level. If acclimatized hops are left in bulk, as heaps, under conditions of low relative humidity then the moisture content of the bracts falls and the cones will tend to disintegrate if they are subsequently manipulated. However, with high relative humidity, (overnight in rainy weather) the moisture content of cones in the outer layer of the heap increases and such cones have a reduced sparkle and there is a change from the original colour after pressing. In contrast, hops correctly dried, acclimatized properly and immediately pressed can be stored for a long time without any change in their quality.

When hops have been dried to the correct moisture content so that any tendence to disintegration is removed they are pressed into large bags known as pockets, to form bales. Portable presses are used for this purpose. The weight of hops in each bale (from which is derived the term "baled hop") averages 60 kg. Hop bales are usually stored on a dry floor in a storehouse equipped with ventilation to allow the access of humid air.

Packing and testing at the hop grower's establishment

Under Middle European conditions the agricultural concern would not be expected to provide any other treatment of the harvested hops than that mentioned above. Different treatments and processing, such as grinding or the production of so-called enriched or granulated hops, require special equipment, a particular technological process and access to laboratory control facilities. These can be done only in establishments which specialize in such work.

Because hops of Czechoslovak provenance are protected by Czechoslovak law, the hop bales must be marked at the place of origin and this is done by the accredited experts of state authorities. Every package leaving the agricultural concern must be sealed and marked with

244

the name of the community of origin and number of the bale. The accompanying weight certificate indicates the gross and net weight.

Weight ratio, and the conversion of fresh to dry hop

Not all agricultural concerns have the same facility *to regulate the volume of hops picked under operating conditions.* Nevertheless, this has to be known during the harvest so that the yield from different hop gardens can be determined, so that the size of the whole harvest can be known and so that the wages to be paid to the operators of harvesting machines can be recorded. Therefore, it is necessary for the volume of fresh as well as of dried hop to be known. The volume of fresh hops can be measured as it leaves the picking machines, but in those agricultural concerns where drying follows immediately after picking, the weighing of the fresh hops interrupts the whole process unless continuous weighing equipment is installed. Where the picking machines are not directly linked to the driers, the weighing operation can be conveniently done and data for the yield of dry hops can be derived from the fresh weight.

The ratio between fresh and dry hop is 4–5:1 depending on the coefficient being used. But this ratio is very approximate and is very variable in individual years.

The agricultural concerns are usually able to determine the moisture content of hops so it is possible to express these relationships more exactly. *A working formula for the conversion of fresh to dried hops is:*

$$Q_2 = Q_1 \frac{100 - \varphi_1}{100 - \varphi_2},$$

where

Q_2 = weight of dried hops,
Q_1 = weight of fresh hops, before drying,
φ_1 = moisture content of fresh hops,
φ_2 = moisture content of hops after drying.

TABLE 56 **Coefficients, used in different years, for the conversion of fresh to dried hops**

Moisture content of dry hop (%)	Moisture content of fresh hop (%)						
	76	77	78	79	80	81	82
5	0.253	0.242	0.232	0.221	0.211	0.200	0.189
6	0.255	0.245	0.234	0.223	0.213	0.202	0.191
7	0.258	0.247	0.237	0.226	0.215	0.204	0.194
8	0.261	0.250	0.239	0.228	0.217	0.207	0.196
9	0.264	0.253	0.242	0.231	0.220	0.209	0.198
10	0.267	0.256	0.244	0.233	0.222	0.211	0.200
11	0.270	0.258	0.247	0.236	0.225	0.213	0.202
12	0.273	0.261	0.250	0.239	0.227	0.216	0.205
13	0.276	0.264	0.253	0.241	0.230	0.213	0.207

For the conversion of dry to fresh hops the following relationship is valid:

$$Q_1 = Q_2 \frac{100 - \varphi_2}{100 - \varphi_1}.$$

245

TABLE 57 Coefficients, used in different years, for the conversion of dried to fresh hops

Moisture content of dry hop (%)	Moisture content of fresh hop (%)						
	76	77	78	79	80	81	82
5	3.958	4.130	4.318	4.523	4.750	5.000	5.277
6	3.917	4.087	4.273	4.476	4.700	4.947	5.222
7	3.875	4.043	4.227	4.429	4.650	4.895	5.167
8	3.833	4.000	4.182	4.381	4.600	4.842	5.111
9	3.792	3.957	4.136	4.333	4.550	4.789	5.056
10	3.750	3.913	4.091	4.286	4.500	4.737	5.000
11	3.708	3.870	4.045	4.238	4.450	4.684	4.944
12	3.667	3.886	4.000	4.190	4.400	4.632	4.889
13	3.625	3.783	3.955	4.143	4.350	4.579	4.833

Harvesting centres

Once the decision had been made as to whether mobile or stationary picking machines should be used, then the development of large-scale production posed new problems concerning driers and post-harvest treatment. Different forms of technology were examined and different types of harvesting centres were designed.

The first type of harvest centre added picking machines to the already existing drier without any real mutual connection. All that was involved was the erection of the various items of harvest equipment at one site. The picking machines were usually of different types according to what was available from home production or by importation. Thus, the harvest was organized in such a way that the operation of the picking machines and the driers was managed separately. Picked hops were transported to driers in bales as had been usual in manual picking. Frequently, machine-picked and manually-picked hops were delivered to the same drier. The only common factor was the output of the equipment. At that time the capacity of the drying equipment was insufficient to cope when the manual and machine picked hops were combined and therefore the machine picked crop was given priority. The machines in use at this time were Bruff D-9; Bruff D-11; ČCH 1 and ČCH 2.

After the introduction of band driers, *the second type of harvest centre* came into being. This type involved the erection of the picking machine and the drier together, but they were still not connected. The output of the picking machines and the band driers was not equal therefore an addition of manually picked hop was required, or the product of separate harvesters went to one band drier. Post-harvest treatment was minimal; hops were

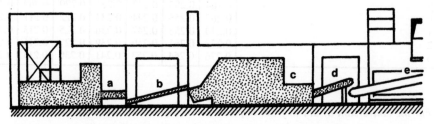

Fig. 116. Diagrammatic representation of a harvesting centre: a – picking machines, b – conveyer of fresh hops, c – band drier, d – transportation of dried hop to acclimatization chamber

246

Fig. 117. Harvest centre.

transported to storehouses and stored in bulk in heaps. This transient type of harvest centre involved picking machines of the following types: Bruff D-11; ČCH 1 and ČCH 2 and band driers PSCH-325 and later, to some extent with Vojvodianka SH 200

The third type was represented by fully-linked machines: the picking machine, the drier and the equipment for post-harvest treatment. The first such centre was erected on the experimental farm of the State Hop Research Institute in SteknÍk near Žatec. Its important feature was the ability to adjust mutually the output of each of the component machines and also to adjust the treatments. Fresh hop bines are fed into one end of this centre and after approximately 8–10 hours the sealed bales of hops come out a the other end. Czechoslovak (type ČCH 3) and Belgian (Allayes M 22, Allayes M 42) picking machines were used. Usually, however, Czechoslovak harvesters ČCH were linked to band driers PCHB-500, PCHB-750 or Vojvodianska SH 200 with acclimatization chambers.

The fourth type consists of modern picking equipment with multi-chamber driers and developed from the first type of centre. Modernization solved the problem of the transport of fresh hops from picking machines to the hopper of the chamber drier, the pneumatic equipment of the drier was modernized as was the mechanical discharge of the individual chambers. The final essential modification was the erection of the acclimatization chamber within the storehouse, built next to the chamber drier, and its link-up with the individual chambers. All of the picking machines in use in Czechoslovakia were assembled and were in this type of centre gradually replaced by more modern versions. The basic criteria have been formulated for presentday harvest centres and take into account the following:

1. location of the centre and its efficiency;
2. the mutual connection of machinery and the basic arrangement of the component parts;
3. technological reliability.

Location of the centre and its efficiency

The location of a harvest centre is governed by a number of factors. But it is generally placed at the centre of a group of hop gardens though outside the village or it can form part of the farm buildings.

The erection of the harvest centres in the midst of a group of hop gardens is technologically very advantageous, because it ensures a short time interval between the cutting down of the plants and their processing and also a short transport distance with the consequent lower transportation costs and use of vehicles. In fact, only a small number of harvest centres is so situated for various reasons including an unsuitable engineering network, difficult communications and lack of accessibility of a water source. Therefore, harvest centres are mostly built within the villages. Under Czechoslovak conditions, where the distance between centre and the hop gardens does not exceed 2 km, this arrangement is technologically satisfactory.

The performance of these harvest centres developed as the performance of their individual components developed. As is shown in Table 54, the performance of current Czechoslovak harvesters cannot cope with more than 20 ha per season. If it is borne in mind that the total number of hop-growing concerns is about 100 and is likely to decrease in the future, then the average area of one concern is more than 100 ha and the most likely area of one hop-growing centre is 50 to 100 ha, then this indicates the required capacity of the harvest centre. It follows that a performace of 20 ha per harvester per season is inadequate.

TABLE 58 **Performance indicators in various forms of hop harvesting (A) two shift and (B) three shift operation (for efficiency group B a new picking machine is ready which will not require so much labour)**

Indicator of performance (efficiency)	Efficiency group	
	A	B
Total efficiency (kg per h of fresh hops)	500	750
Number of working hours per day:		
a) continuous operation	24	24
b) with a time reserve	22	22
Number of shifts per day:		
a) continuous operation	3	3
b) with a time reserve	2	2
Length of harvest (days)	15	15
Planned yield of fresh hop (t)	1.2	1.2
Number of workers on the picking machines:		
a) for the machine including manipulation with waste material	15	10
b) in the hop garden	5	8
c) in transport	2	3
Harvest of fresh hop season (t):		
a) three shift operation	180.0	270.0
b) two shift operation	165.0	247.5
Harvest of dry hop per season (t):		
a) three shift operation	44.5	66.7
b) two shift operation	40.8	61.1
Harvested acreage in ha per season:		
a) three shift operation	36.0	54.0
b) two shift operation	33.0	49.5
Number of workers per ha:		
a) three shift operation	1.83	1.16
b) two shift operation	1.33	0.85
National requirement for labour:		
a) three shift operation	18 300	11 600
b) two shift operation	13 300	8 500
National requirement for of harvest centres:		
a) three shift operation	278	185
b) two shift operation	303	202

As well as the performance of the picking machine, the performance of the other equipment, particularly the drier, also needs to be considered. The most productive type of band drier PCHB-750 dries 750 kg fresh hop per hour and in continuous operation during 16 days of harvest, allowing for an average yield of 1.2 t per ha, is able to process the crop from approximately 50 ha. Thus, it can be seen, that the productivity of driers is higher than of picking machines, but it also indicates the need to build harvest centres with a greater throughput capacity. In general, the harvest centre should match the crop acreage and should be at least able to cope with the crop from 50 ha. It is clear that it should be possible to modify the performance of both the picking machines and the driers to provide higher efficiency.

Interdependence of the component parts of the harvest centre and their basic arrangement

The whole process of mechanized harvest must be continuous, i.e. should be well correlated and not allow the accumulation of hop cones in any one section. In this context following sections need attention: picking, drying, acclimatization and baling.

The interdependence of picking and drying is very difficult to accommodate, it requires a matched efficiency and similar operation rate of both parts. Therefore it is often considered whether these two sections should not be linked to a storage tank to facilitate the treatment of picked hop in a controlled atmosphere. But the efforts made up to the present have not been very successful. This problem does not seem likely to be solved without a straight link-up of the harvester with the drier by conveyor, and this at the same time would facilitate a continuous record of the volume of picked hops.

The drying and conditioning sections are manually linked and this union presents not any technological and administrative problems.

The last section in the harvest centre is concerned with *baling and transportation.* Baling is not a continuous process, but, under the control of the operator, it depends on the accumulation of a volume of hops to be processed. This volume should never exceed the amount to be pressed into one bale. The best technical solution would be for the hops to fall from the conveyer to be gradually compressed into bales. To produce larger stores does not agree with the technical requirements. A certain amount of baled hops can be stored before dispatch and therefore have harvest centres simple storehouses with a capacity of 2–3 day production.

The components of the harvest centres can be linearly or rectangularly arranged, but from the process and operation point of view the best arrangement is linear.

The most important concern in the operation of a harvest centre is *a manually tuned productivity of the picking machine and the drier.* The productivity of the picking machine is very variable because it is affected by many factors including the yield of individual hop gardens, differences in yield from year to year, the maturity of the hop plants, the ratio of cones to other hop plant material, moisture content and fragility of the bines; meteorological conditions, the quality of organization of the harvest and the conditions of the machines. This suggests, that the efficiency of picking machines will vary from day to day during one season as well as in different seasons. Even if those factors dependent on management, such as the organization of the harvest and technical condition of the machines were to be excluded there would still be considerable variations in the efficiency of the picking machines.

The situation concerning the drier is somewhat different, but nevertheless there are factors, which affect its efficiency. Among them the maturity of the hop cones and their moisture content are particularly likely to affect the producivity of the drier. These two factors could be avoided by starting the harvest later, but for many other reasons the start of the harvest cannot be delayed until the cones reach their optimum maturity level, and in any case the maturity reaches the required level during the first quarter of the harvest. Variations in the moisture content of the cones exceed the permissible range (i.e. between 76 and 82 per cent) only very

Fig. 118. Press for the packing of bales of dried and acclimatized hops.

rarely, during particular years. Likewise, the daily variation is small. If surface moisture is concerned, it is easily removed and rarely causes a break in the process longer than 30 minutes. Therefore, the productivity of drying equipment can be regarded as constant, varying only over a very small range which, by comparison with the variability of the picking machines can be neglected.

The efficiency of the whole centre depends on the drier, and this influences the other components. The productivity of the drier should be more than matched by the productivity of the picking machines, so that there is a certain amount of hops in reserve to ensure a continuous and smooth harvesting process even if various unfavourable conditions occurred together. If this were not arranged it would be necessary to regulate the picking machines so that their output would match the working capacity of the drier irrespective of the best rate for picking. Certain problems arise in practice because there may be a need to lower the productivity of the picking machines during the harvest so as to minimize qualitative losses.

Technological reliability

Absolute reliability is a basic condition in the construction of harvest equipment because the quality of the picked hops must remain unsullied. There must, however, always be a question of total reliability where several items of equipment with different features, are working together in one complex. But considerable experience shows the risk to be negligible and it can be accepted that harvest centres are functionally reliable.

Organization of the harvest

The basic process around which the hop harvest should be organized, is the drying operation. Therefore, the preceding operations, both in time and volume, should be matched to the speed of the drier.

The whole organization of real work is aimed at a smooth sequence of operation. Mechanical harvesting starts with pulling down and transporting the hop bines from the garden to a stationary picking machine. Therefore it is necessary, first of all *to calculate the distance travelled by the tractor and trailer in the hop garden in order to pick up a certain number of bines, i.e. the distance-bine ratio.* This distance includes travelling while picking up, and empty return journeys.

If the regularly used tracks are followed, then turns at the ends of rows have little effect on the total distance travelled in loading the trailer and can be ignored. The basis for these calculations is the ratio between the number of bines, the distance between plants in the rows and the distance travelled in loading a trailer.

The time required for pulling down a certain number of bines would depend on the length of the particular rows and the speed of tractor (with turns neglected) and would be calculated according to the formula:

$$t_n = \frac{1}{v} + k_{\mathrm{ú}} + z,$$

where
t_n = time required for loading the trailer (min.);
v = velocity of the tractor with trailer (m.min.$^{-1}$);
$k_{\mathrm{ú}}$ = time required for arrangement of load (min.);
z = time required for stops (min.).

Fig. 119. Hop plants being pulled down for transport to the picking machine.

Once the bines are loaded they are immediately taken to the picking machine.

Time spent in transporting the hops is about half of the total transport time and depends on the distance between the garden and the picking machine and on the speed of the tractor. But the organization of the whole harvesting process has to take into account not only the time spent in transport but also the time lost while the trailer is awaiting its turn at the picking machine and in the hop garden and waiting for space in the roadway from the picking machine to the hop garden, giving the formula:

$$t_d = \frac{2 \cdot L \cdot 60}{V_{str}} + k_1,$$

where

t_t = time taken for the whole transport cycle (min.);
L = distance between hop garden and picking machine (in km);
V_{str} = mean speed of the tractors (k.p.h.);
k_1 = time lost in various delays (min.).

If the work of the transport crew is to be efficiently organized then the time required for pulling down the bines to the trailer should equal the time required for the transport cycle. In this case, if the transport crew is equipped with two tractors, its efficiency should be nearly 100 per cent.

However, one more factor needs to be considered, this is *the time required for the machine to pick the hops from one trailer load*. This depends on the hourly output of the harvester, the number of bines on the trailer and on a coefficient indicating the machine time utilization. This time interval is calculated according to the formula:

252

$$t_{ch} = \frac{n \cdot 60}{W_{ch} \cdot \tau};$$

where

t_{ch} = time required for picking given number of bines (min);
n = number of bines on trailer;
W_{ch} = hourly output of picking machines (number of bines);
τ = coefficient of profit of working time of the picking machine (i.e. ratio of the shift total time to the nett operation time of the machine, e.g. the shift without breakfasts is 460 minutes and the picking machine operated 445 minutes, thus $= \frac{445}{460} = 0.96$).

The operation of the picking machine needs to be smooth, so it is necessary that *the time required for pulling down*, a given number of bines equals *the time required for the transport cycle plus the time required for picking the load of the trailer,* thus:

$$t_n + t_d = t_{ch},$$

where

t_n = time required for loading a trailer;
t_d = time of a transport cycle;
t_{ch} = time required for picking the trailer load;

or provided:

$$t_n + t_d = t_c,$$

where

$t_c = t_{ch},$
t_c = time required for getting the trailer to the picking machine.
If an equation is formed, using the expressions derived earlier, it will appear as follows:

$$\frac{1}{v} + k_{\acute{u}} + z + \frac{2 \cdot L \cdot 60}{V_{st\check{r}}} + k_1 = \frac{n \cdot 60}{W_{ch} \cdot \tau};$$

If during a working shift both sides of this equation are equal, then the correct proportion of time and capacity involved in the pulling down of the bines, the transport and the picking operation will have been achieved. But the relationship can be affected by the following:

$$t_c \gtrless t_{ch}.$$

If $t_c > t_{ch}$ and at the same time $t_d > t_n$, then it will be necessary to provide more tractors to the transport crew, but if $t_c < t_{ch}$ and at the same time $t_n > t_d$, then it will be necessary to increase the speed of pulling down the bines, and if $t_c < t_{ch}$, then it will be necessary to take out one tractor (if two are being used for transport) or to increase the output of the picking machine.

These various relationships are shown here so that they can be plotted graphically after the substitution of real values as encountered in local operations. This facilitates the necessary organizational modifications when the weakness of individual elements is found, so that suitable changes can be made to individual operations.

The evaluation of the quality of hop purchased from growers

The purchase of hop of the growers is governed in Czechoslovakia by two special state standards (CSN 46 2510 – Hop and CSN 46 2520 – Hop tests). Both standards were innovated and their last modification is valid since the 1st August 1986.

There are two categories of characteristics used for evaluation of hop purchased from growers, namely

1. *objective characteristics* investigated by objective methods in laboratories, among which it has to be mentioned: damage to cones by harmful factors, the rate of knocks (damages to the colour of cones), the rate of disintegrated cones, the rate of biological adulterants, the moisture content of cones and the content of α-bitter acids (conductometrical value),

2. *subjective characteristic* evaluated by senses – biological growth, the structure of spindle, the scent, the presence of seeds (negative characteristic), the colour of cones, the colour of lupuline.

There are four classes of quality. The State Inspection of Quality of crops and products of food industry establishes for every quality class the samples of typical quality. These samples are made known to every purchasing centre before the purchase starts. The representatives of the grower and of the purchasing centre decide about the quality class of the hop which is purchased. The results of laboratory analyses of the purchased hop as well as the typical samples are available to the experts which decide about the quality class given to the purchased hop.

Generally, three levels of quality are distinguished. On the first level, the deviations from the ideal parameters can be tolerated. On the second, these are over limits and the price should be reduced. In the third case, some corrective measures are required to make the product acceptable for commerce. These measures are naturally to the account of the grower. So the rate of biological adulterants is limited to 3 %. When their quantity ranges from 3 to 7 % the price is reduced, when it exceeds this limit, the product has to be recleaned on the grower's costs. The content of alpha bitter acids is limited with 4 %. Every 0.5 % over or below this value are reflected by positive or negative charges respectively.

The purchased hop is delivered to the commercial organization in marked (place of origin, weight and number of the bale) and sealed bales. Under the presence of the representative of grower, the samples are taken from every bale. The mixture of them serves as an average. The average sample is divided in three parts, one for laboratories, the second for price estimation and the third for eventual claims.

The arbitrary organ for the claims is the State Quality Inspection of crops and products of food industry. The claims have to be decided during two weeks and the decision is definitive.

CHAPTER IV
THE ORGANIZATION AND ECONOMICS
OF HOP PRODUCTION

Hop production demands increasing attention from the managers of the agricultural enterprises, from hop growers organizations and from all institutions engaged in this branch of agriculture. Therefore it is not only necessary to pay attention to the biological and technological questions discussed in the preceding chapters, but also to those economic factors which concern the application of labour in the production process. The application of new technology and new means of production made it possible to concentrate hop production and make it more specialized. These changes required new methods of organization and management. It is evident that the use of more precise methods in organization and management is increasing in large concerns as well as in individual production units.

PRESENT STATE AND FUTURE TRENDS IN THE USE
OF LABOUR IN HOP PLANTATIONS

Developments in hop production involved a number of quantitative and qualitative changes:
1. the number of qualified workers increased as it did in agriculture in general;
2. new technological processes were introduced and productivity increased;
3. more productive and more efficient equipment continues to be introduced for all main operations;
4. production became concentrated in particular areas and workers in the industry became specialists.

These factors in the development of hop-growing formed the basis for new forms of organization and management and also provided conditions for the further intensification of the production process. This has been demonstrated by the long-term development of certain intensifying factors which have increased the yield of hops. The study of correlations also shows that productivity increased only when all of the factors investigated in hop research were acting together.

Yields gradually increased during the years 1961–1975 as the decisive intensification factors were increasingly applied. Apart from years of crop failure, yields maintained a growing trend during both of the five year plans. During the last five year period, the yields were stable at 1.1 to 1.2 t.ha^{-1} for comparison, the yields before 1960 were 0.7 to 0.8 t.ha^{-1}. Thus, during the last 15 years hop productivity has significantly increased and yields have become stabilized at a substantially higher level. This change came as a consequence of the whole set of intensifying factors having been applied more and more strictly since the beginning of the third five year plan. Scientific and advanced knowledge, suitably applied in practice, contributed to this important phenomenon and became one of the most important production forces.

It was proved that increased yield is a function of the whole complex of simultaneously and homogeneously applied intensifying factors and that yield, as a dependent variable, depends

TABLE 59   Mechanical equipment and other technical items delivered for use in hop
            production in the years 1971 – 1978

Equipment or construction	1971	1972	1973	1974	1975	1976	1977	1978 plan
Picking machines:								
ČCH-4	36	27	28	44	–	–	–	–
LČCH-1	–	–	–	–	39	24	20	35
Reconstructed driers KSCH:								
Number of reconstructed								
chambers	85	67	77	64	55	3	4	–
Rec. PSCH 325	1	2	3	4	3	2	–	1
PCHB 750 K	6	5	3	7	13	16	4	11
Self propelled								
pesticide sprayers:								
Myers[x]	30	–	–	13	–	–	–	–
Kertitox[x]	–	–	–	–	30	50	35	32
Erection of wire-work								
structures in hop gardens:								
area (ha)	297	412	585	625	812	823	683	734
investments, in Czechoslovak								
crowns (1000 s)	17.765	24.844	36.485	40.072	50.161	51.300	42.408	44.257

[x] Myers and Kertitox – trade marks of sprayers.

on a complex of independent variables. These include plant spacing, cultivated varieties, intensive manuring, specialized production and level of mechanization. This understanding is most important for planning the further intensification of the production process. Therefore it will always be necessary to apply up-to-date intensifying factors, as a whole complex, and with

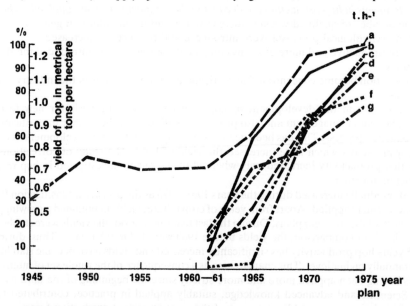

Fig. 120. Effect of various factors on increases in yield of hops over a period of 30 years: a – yield per ha, b – improved varieties, c – rate of manure applications, d – plant spacing, e – mechanical spring cut, f – specialization in hop-growing, g – mechanical harvesting.

a balanced correlation in order to stabilize the yield and to assemble all productive reserves for future yield increases.

Moreover, it will be necessary to increase the rate of application of new important intensification factors, such as irrigation, which has been adequately used only since 1975. It has been shown that irrigation can double the yield in critically dry years and that its average effect over dry and wet years is to increase the yield by some 25 per cent. This suggests that in terms of crop management this factor is still not fully exploited. Depending on the scale of its future application a further growth of yield can be expected between 1985 and 1990 and this growth could bring the total yield up to $1.3-1.4$ t.ha^{-1} nationwide and in some localities, with almost 100 per cent irrigation it could reach 1.5 t.ha^{-1}. It has to be understood, however, that all of the other intensification factors have to be applied together with this new factor of irrigation and it is necessary to ensure that other aspects of production are of a high standard. They are based on management, total care of the stand, cultural practices (reducing the number of missing plants, improved care for dislodged bines, establishment of the hop garden in the autumn using container-grown stocks, harvest in the first year after planting, intensive and reasonable nutrient applications, early and effective protection against diseases and pests, and reducing harvest losses). The quality of harvested hops is heavily dependent on harvest and postharvest treatment.

It can be concluded that the present level of hop production could be increased and that the first task is to make use of the existing reserves. On the other hand, it is clear, that contemporary hop production is a technologically, technically and administratively well-developed branch of plant production, where the advantages of large-scale agricultural production are apparent and indicate ways in which other branches of agriculture could be developed.

CONSUMPTION AND USAGE OF LABOUR AND EQUIPMENT IN HOP PRODUCTION

Although there is an increased use of mechanization in hop-growing, the requirements for labour still remain very high. This is due to the properties of the hop plant itself (perennial rootstock with an annual climbing bine) and to the requirements for its environment if its growth and development are to be successful. Some operations, or parts of them, require more human involvement either due to imperfections in the available equipment or due to a high requirement for certain procedures which are very labour-intensive.

Any estimate and evaluation of the required labour will be based on the following data:
1. total area of fertile hop gardens;
2. selected agricultural procedure;
3. technical and operational parameters of the equipment;
4. time limits of agrotechnical works (calendar dead-lines for certain operations assuring the hop production for the whole vegetative period of a fertile hop-garden),
5. local geography and pedology.

The calculation is based on the so-called exerted work rate, i.e. on the total work required, related to the area (hectare). The exerted work rate may be direct work or extended work. The direct exerted work rate is the time required for particular operations done immediately in the hop garden, but experience has shown that extended exerted work rate is decisive. This relates to the total required time and so includes work not immediately effected in the hop garden but which is required as part of the production process. The extended exerted work rate thus includes such extra as time spent on the daily maintenance of tractors and equipment and the time required for travel to and from the hop garden. The distance of the hop-growing area from the point where the machinery is concentrated is taken as an average of 3 km. Mean length of a single traverse in the hop garden is taken as 250 to 300 m. Deviations from these values can be assumed to be negligible in calculations because they are mutually balanced.

The standards for exerted work rates include all categories of time required for productive duties; these include actual working time and time involved in preparation for work.

Preparation time intervals include:
(a) for mechanized operations
  resetting utensil into working order,
  travel from row to row,
  travel to and from the hop garden,
  cleaning equipment,
  preparing it for transport to and from the garden;
(b) for manual operations
  putting on working clothes and preparing tools,
  walking to and from working area,
  cleaning utensil and clothing (e.g. muddy boots).
Working intervals include:
(a) for mechanized operations
  handing over and doing the work,
  personal need and rest,
  necessary discussion of work,
  quality check,
  refreshment break;
(b) for manual operations
  handing over instructions and doing the work,
  personal need and rest,
  discussion of work,
  refreshment break.

The exerted work rate is calculated according to the planting pattern of the hop garden. With so-called broad spacing the average distance between rows is 280 cm. If certain operations need to be distinguished because they are in pylon rows or in normal rows, then the ratio of these rows for narrow spacing can be taken as 1:4 and for broad spacing as 1:2.

Before the calculation of the exerted work rate, it may be necessary to break down the whole production process into its component operations. Several other operations, however, are combined (e.g. manuring, cultivation, spraying and transport of materials). Therefore it is neither necessary nor satisfactory to plot hop production chronologically but according to the various homogeneous and specific operational complexes.

For determining the likely exerted work rate in a particular situation and so as to determine the amount of labour and machinery required, it is usual to make use of norms or standards (basic or derived), established by statistically justified calculation in the Institute for agricultural work and food industry. These values have been established for hops in following conditions: medium and heavy soil, slope up to 5°, narrow spacing 150 cm and broad spacing of 280 cm.

Requirement of labour and equipment during one growing season

The number of man hours required depends in part, on the operational plan to be used with the particular plantation pattern, and in part on the mechanical equipment and its efficiency. For the calculation of man hours required standard data for exerted work rates in hop production can be used or the figures can be derived from the known efficiency of the various item of equipment.

Assessment of work required per ha for hop production:
Conditions:
broad (280 cm) spacing of rows,
heavy soil,
planned yield 1.2 t.ha^{-1},
ČCH 3 type harvester.
Individual operations:
1. Clearing the hop garden,
cutting and burning bines,
taking out the hooks,
manual removal, loading and unloading of damaged wire work;
work required:
tractor driver – 2.66 hours,
other operators – 52.47 hours.
2. Post harvest soil processing:
harrowing using of Z-40 harrow,
ploughing with PSON 2 – Z-50 plough;
work required:
tractor driver – 4.19 hours,
other operators – 2.66 hours.
3. Manuring with 30 t.ha^{-1} farmyard manure:
loading of dung with NuJN-100,
weighing and transport with RM,
spreading with RM;
work required:
tractor driver – 13.70 hours,
other operators – O hours.
4. Soil cultivations and spring cut of hops:
harrowing (twice), ploughing and cutting with OŘCH 1.2 – Z-50,
digging out and cutting the remnants of mechanized cut of hop,
levelling the ridges with Z-50;
work required:
tractor driver – 9.70 hours,
other operators – 12.40 hours.
5. Suspending the training wire:
loading, transport and unloading the material,
suspending and fixing the wire rope;
work required:
tractor driver – 0.05 hour,
other operators – 148.70 hours.
6. Treatment during the growing season:
training after the cut,
cleaning before ploughing in,
ploughing in twice with PSON 2 – Z-40,
inter-row cultivation (5 times) with Z-40,
training displaced bines,
suspending dislodged plants – up to 5 per cent;
work required:
tractor driver – 15.36 hours,
other operators – 210.12 hours.
7. Crop protection:

spraying by aircraft,
irrigation;
work required:
tractor driver – 16.14 hours,
other operators – 4.48 hours.
8. Manuring and fertilizer application:
manuring prior to ploughing,
additional manuring before first ploughing-in,
additional fertilizing prior to blossoming;
work required:
tractor driver – 0.73 hour,
other operators – 17.80 hours.
9. Harvest and after-harvest treatment:
cutting and pulling down the bines onto the trailer,
transporting the bines to picking machine Z-3011,
picking the hops – ČCH 3,
manipulation of green hops,
removal of waste material,
hop drying – (louvre drier),
baling the hops;
work required:
manipulation with machines – 71.6 hours,
other operators – 601.7 hours,
   E x p l a n a t i o n   o f   c o m m e r c i a l l y   u s e d   m a r k s   o f   m e c h a n i z a t i o n
e q u i p m e n t :

NuJN-100	= multi-purpose crane loader,
RM	= dung spreader,
Z-40	= tractor ZETOR with power 40 kW,
PSON 2	= plough with two shares for works in hop gardens,
Z-50	= tractor ZETOR with power 50 kW,
ČCH-3	= hop harvester from 3rd series of production,
OŘCH 1, OŘCH 2	= hop cutter,
Z-3011	= tractor ZETOR with power 30 kW.

Total man hours of work required to produce finished hops from one hectare of hop garden:
tractor drivers – 134.13 hours.
other operators – 1050.48 hours.

This total of 1184 hours is evenly spread over the season. Half of it is involved in the harvest and post-harvest treatments and one third is used in crop protection, manuring and suspending the training wire. If manual fixing of the wire were to be replaced by a mechanized procedure, then all workers except those involved in the harvest will be permanent employees of their respective concern and this will contribute to the stabilization of hop-growing areas within the framework of agriculture as a whole. If irrigation is required then the amount of labour involved depends on the irrigation system being used.

ORGANIZATION AND MANAGEMENT OF HOP-GROWING UNITS

The social division of work which has developed through the evolution of productive forces and from new knowledge resulting from scientific and technical progress, has caused individual branches of production to become independent not only in factories but also in agriculture. This can be demonstrated by the specialization in hop-growing already developed by the research institutions.

The number of other production branches on the form decreased during this progressive specialization in hop production, which thus became intensive and large-scale. Nowadays only such other farming activities are retained in hop-growing concerns as contribute to the necessary balance between the biological requirements (e.g. cattle, to produce farmyard manure) or which cannot be excluded (e.g. production of cereals or sugar beet).

Further developments in productivity and increased scientific and technical progress could lead to the total independence of hop production i.e. its complete separation from other farming activities. This would involve a completely new approach to the management of hop production.

Different forms of management in hop production

In general the management form which best meets the requirements of the further development of *large-scale hop production, is direct management without intermediaries*, i.e. the direct management of the production of this one crop. This functional type of management structure allows the full realization of specialization in the various individual functions and thus encourages management by qualified experts – hop-growing agronomists. These will be in charge of the specific means of production and the labour force and so will have their own authority and responsibility. In terms of organization and administration the management will be made up of special, i.e. hop-growing, production units, which will work on their own account with their own final profit or loss.

With the gradual introduction of new forms of management, and different form of cooperation resulting from the specilization of production, two variants of direct management of hop production can be considered.

First variant: hop-growing groups

Two possibilities can be distinguished here:

A management system in which hop-growing is not yet a fully integrated part of the agricultural unit. In this system a central body hires to the cooperating concerns special and heavy-duty, technical equipment, which could not be afforded nor fully used by individual members of the cooperative group. Other minor operations, mainly manual, as well as harvest are performed by the concerns themselves. In this system it is best, if the cooperating centre is managed by a hop-growing agronomist who can manage the special heavy-duty equipment (machines for hop cutting, applying protectants and fertilizers and for planting) and can organize his collegues in the individual cooperating concerns. He arranges the time table of work on the season and having been provided with the necessary means of production he is personally responsible for the production. Such managers are paid according to their production results.

In the second system of full cooperation, i.e. with all the marketable products of the concern included within the one organization it is reasonable to build into the framework an independent hop-growing production and administrative unit (see next variant).

Second variant: organization of production and administrative units

It is clear, *that production and administrative units can be constituted for concerns, or cooperating groups, of different dimensions and therefore we can consider here:*

I. production and administrative units (PAU) for a hop-growing area of 100–200 ha,
II. PAU for a hop-growing area of approximately 500 ha.

The management structure of these categories differs in the following ways:

Category I. the technical staff responsible for hop production includes –

manager of production and of the administrative unit – qualified, with an agricultural university degree,

1 technician hop grower – mechanical specialist qualified at a specialized college,

1–2 junior technicians,

an accounts clerk,

storesman.

Category II. the size and structure of the management and technical staffs depends on the overall size of the hop-growing area, but for 500 ha would be –

1. management –

unit manager,

independent agronomist technician – hop grower,

independent machinery specialist – hop grower,

independent technician,

economist,

1–2 bookkeepers,

administrative worker.

2. central group –

mechanization group – provides all heavy-duty mechanical services, including crop protection, manuring and composting as well as a group for the erection of hop garden structure and repairs to the wirework.

3. organizing departments –

these departments provide the hop production on the area allotted to them in the framework of the farm,

the make-up of this group depends on the configuration of the terrain, on the site of production and on administrative unit, on the communication network and on future erection. These groups should service approximately 100 ha but a minimum of 50 ha and should be managed by one or two independent agronomist-technicians and one bookkeeper according to the serviced area. They should have at their disposal a certain number of operators in permanent employment as tractor drivers, workers etc.

This new system of production management, called branch management, has come to be recommended after considerable experience and the gaining of new knowledge. It makes use of scientific and technical progress as well as qualified management, it makes best use of mechanical equipment and reduces the labour requirement, even in administrative departments. Hop gardens will no longer be controlled from centres which are managing a number of branches, but will be in an independent production and administrative unit (PAU – hop). Depending on its size this unit will be allocated an appropriate number of permanent and seasonal workers as well as temporary workers specially for spring work in the garden (suspending the training wire, training) and for harvest. Furthermore, it will be allocated all the mechanized equipment required for hop production (e.g. tractors, sprayers, fertilizer and FYM spreaders and picking machines) and buildings (e.g. driers, housing for temporary workers and implement sheds) and such other equipment as is needed for hop growing. The permanent and the temporary workers will be organized from the central production and administrative unit (central duties) as well as within their individual sections.

The total area of hop gardens will be divided, according to their individual administrative units, into sections which will be managed by specialist independent agronomists and agronomist-technician hop growers. The labour required for hop-growing will be provided in these sections by permanent and temporary workers. But the cutting operation, crop protection, manuring with FYM and artificial fertilizers, erection of new hop gardens, and the maintenance and repairs of existing gardens will be provided by the production and administrative unit which has the required workers at its disposal. Every section will be allocated the necessary hop-growing equipment. At the same time these sections will be responsible for the picking machines, driers, housing facilities and other operational buildings entrusted to them. Independent agronomist-technician hop growers will oversee all of the work of the central production and administrative unit.

This type of management can make full use of the latest technical and scientific advances, because the fully qualified staff works in only one branch and can therefore proceed in a fully qualified manner.

ECONOMIC ANALYSES OF HOP PRODUCTION

Economic analyses facilitate decisions about planned production, because it provides information on the productivity and economy of the enterprise and thus has an important function in management. It provides information on interrelationships and on factors affecting production and consequently aids decisions on technical, technological, administrative and economic measures. Its various features provide information on such phases of management as planning, organizing and operational management.

The information provided by economic analysis is either quantitative or qualitative.

The basic information is as quantitative data, i.e. the real returns and costs of planned production. Qualitative data are in the main derived from quantitative data and represent generalizations or express, in general terms, the social side of production.

An economic analysis can include the following items:
joint-activities within the enterprise,
the results and correlations of joint-activities,
overall economics of the enterprise,
understanding of the social aspects of the enterprise.

Depending on their action radius economic analyses are either partial or summarizing.

Partial analyses examine only selected phases of production, involving individual units within the organisation.

Summarizing analyses investigate all parts of the enterprise, i.e. all branches of production and organization.

Economic analyses can be further classified as:
deep and orientation analyses,
long-term and short-term analyses,
periodical and non-periodical analyses (regularly and randomly repeated),
documented records,
process controls.

Economic analyses are subjected to generally valid principles including:
complexity of analysis,
definition of main point,
objectivity of the analysis,
collectivity of the analysis,
reliability and effectiveness of the analysis.

Economic analyses apply certain basic and specific methods of investigation involving a comparative method, analysis and synthesis.

The comparative method is the most widely used in economic analysis, especially in the analysis of orientation. For the application of this method, the bases of comparison are most decisive and the phenomena and signs to be compared are selected from them.

Comparisons can be made arithmetically,
using the difference between reality and the selected basis,
using the rate of reality to basis.

The analyses can use the following items for the comparative basis:
results obtained in previous periods,
unusual (above or below the average) results,
optimum construction.

Comparisons can be made in *a horizontal direction* (with other enterprises or branches in a given time interval) or in vertical direction (in the same enterprise or branch at different times).

All of these principles and possibilities of economic analysis can be used in hop production. Within a particular concern it is important to relate the level and efficiency of material and labour investments to the quantity and quality of the product, the main aim being to examine the return from individual investment items. However, with a biological product it is necessary to know the conditions under which the investments were made, if the conclusions are to be justified. Thus, very often in practice, there are cases in which a higher level of investment (namely for labour) was unavoidable because of poor natural conditions. Here an increased input did not increase production, but prevented its fall. In general it can be said, that increased investment in hop production is usually caused, and consumed, by a deterioration in natural conditions.

Another phenomenon to be examined is the relationship between the amount of human endeavour applied to production and the value of the product. Likewise the problems of organization and management of hop production by individual production and administrative units, as well as in the whole enterprise, can become the subject of analysis.

New trends in management necessitate complex economic analyses of the whole branch, with the aim of determining the quality and quantity of production as well as the factors which influence these parameters. Where an economic analysis notes the function of these factors over a longer period, it can provide useful information for further growth of productivity.

The analysis of hop production must lead to the adaption of measures to eliminate such faults as the analysis has revealed. This mainly concerns re-planting of missing plants, the application of suitable means of crop protection, the reduction of harvest losses to a minimum, and the use of such drying and acclimatizing equipment as will maintain the quality of hop cones.

CHAPTER V

RATIONALIZATION AND SCIENTIFIC AND TECHNICAL DEVELOPMENTS APPLIED TO THE PRODUCTION OF HOPS

The concept of rationalization implies a systematic rational arrangement of operational procedures to provide the maximum productivity at the minimum expense. Complex rationalization is a means of ensuring a high level of economic efficiency in production. It directs the development of scientific and technical progress and becomes an integral organic part of the daily work in hop-growing.

COMPLEX RATIONALIZATION

Rationalization is based on the knowledge of those factors which affect the growth of productivity by labour, and the intensity of production; it has the following specific features:
– the working man is both its object and its subject;
– its results benefit all working people;
– rationalization measures must not be detrimental to the safety and health of the workers involved;
– it concerns the human aspects of work, i.e. it makes the work process physically less difficult and less stressful;
– it is flexible in that it can continue to be improved.
The complexity of rationalization involves the application of the conceived system of production and includes the following fields:
– production,
– management, turn and work.
Rationalization is a specific constituent of the complex socialist policy because it pervades all sections of work and largely controls it.
Rationalization is complex, and can be classified as:
– external, (from the viewpoint of the respective agricultural concern) involving the processes of concentration, centralization, specialization and cooperation;
– internal, covering the biological, technical and administrative rationalization.
With regard to the wide-scale acceptance of rationalization the measures are distinguished as:
– short-term (regarded as non-investment) and
– long-term (involving investment).
Short-term rationalization measures are those which do not require the construction of new buildings or the purchase of mechanical equipment, and mainly represent changes in working methods as found in so-called work studies. Such studies can indicate new work procedures which will achieve the given aim with a lower consumption of working time and a lower energy input. This system of rationalization is based on the fact that every working procedure can be executed by another more efficient method and the task of rationalization is to find this

method. In the short-term the value of these measures relies on their immediate positive effect on the production process.

Recently, the question of rationalization has been solved mainly by means of *methods involving the time to be taken for various jobs being fixed in advance,* therefore so-called *time-and-motion studies* are most frequent. These are based on the analysis of working and non-working movements, on the comparison of the speed of certain time movements with standard speeds and on composing a new working method with its reduced requirement for working time. It is, however, necessary to combine this analysis and synthesis of working acts with a respect for the human involvement in the work, i.e. for the modification of the working environment, especially the physical factors of this environment. This method is mainly used for manual work involving repeated operations.

Other operations can be investigated by so-called work studies (see diagram in Table 60).

TABLE 60   **Diagram of a work study programme**

Stages		
Studies of working methods	selection of the problem	measurements
Symbolic or graphical method	entry of working time shot procedures	time pattern
The sequence of questions	critical test	analysis of time
New working method	design	calculation of time standards
	verification of the design	application of standards
	introduction	
	into the production	
	process	

Most important is the so-called critical test, the purpose of which is to evaluate recent operations and to draft variants based on these evaluations. It is based on a logical sequence of basic operations leading to a better performance of selected operations (see Table 61).

The design of a new method is based on the selection of the variants which are more suitable in the given conditions and for the aims of the operation.

Long-term rationalization measures are those which will give a longer return on such investments as those in the construction of new production capacity, the purchase of new machinery and equipment and from the introduction of new technology. This form of rationalization is relatively easy but is costly. Therefore careful consideration must be given as to whether the proposed operations are suitable in relation to the long-term aims of production.

Main aims in the rationalization of hop production can be listed, according to the general aims of agriculture in following spheres:

1. biological sphere,
2. technical sphere,
3. economic and administrative sphere,
4. sphere of labour input.

266

TABLE 61 **Schedule of a critical test – the example used here is the taking down of bines for harvest**

Question	Recent method	Alternative method	Selection
What happens?	pulling down the bines and loading them on to the trailer	– pull down – not to pull down – not to load – not to adjust – better adjust	pull down
When?	at the best time for feeding the picking machine, prior to picking, during passage through the hop garden	– in the period before picking – for putting into reserve – during passage through the hop garden – during stopping time	during passage through the hop garden
Where?	from the platform of the trailer used for transporting the bines	– from a different construction on the trailer – from the trailer – from a special separate platform – from the ground	from a different trailer
How?	by grasping, stretching and tearing the bine and adjusting it on the trailer	– to stretch the bines – to fix the bines – to pull the bines down and not to adjust – to pull the bines down mechanically	to pull the bines down mechanically
Who?	tractor driver and a group of 4-6 workers	– no worker – tractor driver – tractor driver plus two assistants – tractor driver plus more workers	tractor driver plus two workers under bad conditions

Rationalization measures are mutually linked and conditioned between the different spheres. The biological sphere can be taken as the base, because measures taken there will affect not only production technology, but also the economics of the operation including the organization of labour.

The biological sphere primarily concerns the breeding of varieties of hop, to give higher yield of cones with a greater content of bitter components. The basic requirement of such rationalization is to preserve the specific character of the original hop from Žatec with regard to the structure of the cones and the bitter substance components. Agriculturally, rationalization primarily involves the modification of the habit of the hop plant to give an increased ratio of cones to the other parts of the plant.

In the technical sphere, rationalization will investigate and develop those items of equipment which improve on natural processes. This may lead to further important progress in the industrialization of hop production. Among the rationalization measures which have led to a greater intensity of hop production are the introduction of new powerful irrigation systems and various innovations in the machinery, machines and compounds used in crop protection. The function of rationalization at harvest is to improve the speed of picking machines and to lower the losses during picking. A specific task is here to determine how to transport whole bines, or their parts, to the stationary harvesting lines in order to increase the mass of transported cones per km and to reduce the man hours involved in this process.

In the economic and administrative sphere rationalization aims to apply automated management systems to operations, production management and filling. This system should lead to a controlled production process in which subjective decisions of individuals will be made objective by the intervention of computing techniques.

As to the sphere of labour input, rationalization depends on the results obtained in the technical sphere. The main intention here is to provide better working conditions. This is not concerned only with the presence of the operators in their jobs but also such aspects as the modification of the environment and the introduction of new work stimuli. Here, it has to be remembered that hop production is very intensive and will always need more labour than is required by less intensive crops.

Those agricultural concerns permanently engaged in questions of complex socialist rationalization work towards those conditions which lead to efficient hop production and to a generally successful economy.

PRINCIPLES OF PLANNING FOR THE FUTURE DEVELOPMENT OF HOP PRODUCTION

Because hop culture is permanent and has a high requirement for investment, constant attention must be given not only to annual production plans but primarily to planning perspectives which should serve as a basis for the annual plans. This has become increasingly important in the recent reconstruction of Czechoslovak hop production, based on large-scale agricultural production policies as well as or its modernization. The introduction of new technology and technical equipment as well as other modernization measures necessarily involves increased investment and, therefore, planning has to cover long time intervals, usually 5–10 years and in some cases (e.g. for costly harvest lines) even longer periods. If these plans reach fruition in the respective years then the overall aim will have been achieved and the whole branch can then be reconstructed using modern large-scale production technology. The return on investments will usually be after more than 2–3 years.

This long-term planning of hop production is also, in part, necessitated by the need to provide other localities for hop-growing. As hops are a perennial crop, a change over of cropping area can be arranged only once a year, whereas large-scale planning, on a district or regional basis has to allow for the natural cessation of production in unsuitable hop-growing areas and for the enlargement of hop gardens in developing areas. Such a process of gradual concentration of hop growing in the most suitable localities can be achieved only by broadly-based planning covering 15–20 year periods. Thus, a gradual and optimized concentration of hop-growing into one concern or one locality will make best possible use of existing as well as future mechanization. In the past an area of approximately 30 ha was regarded as the minimum workable area, but now it is about 50 ha. This newer minimum area is based on the capability of the second generation machines, particularly for harvest lines, which can process the output of 50 ha per season under conditions of uninterrupted operation and a well-organized work force.

The maximum concentration in one locality should be approximately 100 ha, because one responsible director cannot effectively manage the production of a larger area, especially if the hop gardens are in different localities. At present, even bigger organization units exist and will continue, but such large complexes should be under a management team of agronomists according to the principles of product management (see page 261).

As mentioned above planning for the development of hop production should cover a period of 15–20 years for those organization units that are bigger than a single concern or for a larger economic complex because of the essential slowness of concentration and the allocation of expensive equipment, particularly harvest lines. It needs to be borne in mind, that the estimated value of a continually-working second generation harvest line can be as much as 10 millions CKR (Czech crowns). This is sufficient argument for the long-term planning of hop production especially at the district or regional level, where the volume of investment can reach hundreds of millions CKR. Therefore the conception plans must justify the likely investment and demonstrate their economic suitability and efficiency before the programmes for years can be started.

These conceptions of the development of hop-growing can be distinguished, according to their scope, as developing, consolidating or a combination of developing and consolidating. Their aim is evident from their names, but it will be useful to mention their main features.

The developing conception of hop production is usually elaborated for the needs of the particular agricultural concern and in order for superior authorities to decide about the further development of hop production, i.e. to increase the volume and quality by an increase in yield per unit area and by management innovation and an increase in the hop-planted areas. This type of conception involves those hop-growing concerns which are already consolidated in a group, whose further existence, and the growth of their production, the solution of problems of concentration, specialization, and application of large-scale technologies is already included in general plans.

This involves large investments and changes in the location of hop-growing areas and, hence, a well-elaborated conception for a period of 15–20 years becomes very important. The same reasoning applies to those concerns where the concentration of hop-growing areas has been fully achieved.

The consolidating conception of hop production is recommended for those concerns, or their constituent parts, where hop production has decreased to below the average of the district, region group of neighbouring concerns. The aim of this form of planning is to determine the reasons for the deterioration and to propose measures for the stabilization of production at the required level. The realization of such measures takes, on average, 5 years.

The combined (developing and consolidating) conception of hop production is a synthesis of the two preceding types, in the sense that the developing conception is based on consolidation. This form of planning mainly relates to individual concerns or their parts, a larger scale situation being considered only exceptionally. This usually involves concerns where production has considerably decreased and therefore its redevelopment has to be given attention in two periods: in the first period, that of consolidation, the production must reach a level equal to that of the district or region, and in the second, or development period, hop production is re-established on this basis. Each of these periods usually lasts 5 years, so that the whole combined conception develops over at least 10 years.

The type of conception can also be classified, according to the area involved, as internal (one concern only) or external (district, region, territory, republic). The scope of such categories is evident from their names, but they also differ in the period over which they usually extend and in certain obligations concerning investment. These distinctions will become clearer as the respective methodology is discussed.

Fig. 121. A modern harvest centre.

Experience has shown that the development and analysis of any conceptual planning involve the following points:
a) introduction,
b) analysis of the level of production reached during the previous 3–5 years,
c) survey of main existing agrotechnical, technological, technical and economic measures,
d) likely economic and administrative problems of the proposed conception,
e) conclusion,
f) enclosures (e.g. tables, diagrams, maps of hop blocks and statements from higher authorities).

Fig. 122. Buying centre of hop cones.

The introduction briefly describes the scope and aim of the conception, its type and the period for which is planned and the authority which demanded its preparation.

The analysis of the level of production reached during the previous 3–5 years should provide a study of the development of hop production reached hitherto. The aim is to determine the main positive and negative features as well as the scope for increased production as a basis for the re-assessment of production trends. This analysis is also necessary in order to define the main technical and administrative measures required for increased production and to provide economic indicators for the future.

The survey of main agrotechnical, technological, technical and economic measures is the main feature of the whole concept. On the basis of the foregoing analysis this survey primarily determines the development of hop-growing areas and their expected yield per hectare taking into consideration the capability of the individual concerns as well as target figures provided by superior authority. In the development of hop-growing areas, the most important requirement is a time-table for the creation of new hop gardens and abolition of old ones and to provide costings as a basis for planned investment. Furthermore it should facilitate an increase in yield per ha and the optimum utilization of new technical equipment, by:

placing new hop gardens in the best pedological and territorial conditions;

giving priority to hops on the most suitable land and in proximity to harvest centres,

providing the maximum utilization of land with the existing, or future, irrigation,

selection of those features, particularly the slope, of the ground so as to utilize, fully, all mobile mechanical equipment,

making the maximum concentration of hop-growing areas into blocks with a surface area of not less than 6 ha.

Experience, mainly in bigger hop-growing concerns has shown that it is very useful to prepare a plan to show the proposed new areas for the extension of existing gardens and the overall concentration of hop-growing unit, with an indication of the present stage and the possible future layout of irrigation systems over the whole concern as well as in the individual production units.

This classification of perspectives is important for decision-making about further intensification of hop production.

Based on the prepared analysis the next part of the survey indicates the main measures required for the further improvement of agrotechnical methods, nutrition, protection, large-scale production technology and harvesting. The harvest of hop needs to be discussed separately, because not only does it concern technology, but it also involves large investments. Finally, the plan will enumerate those measures essential for success. These include the time schedule and the responsibilities of individuals and will define those problems which should be decided in cooperation with superior authorities.

Economic and administrative problems of the proposed conception define the total volume of investments for the whole period and for individual years and thus provide a basis and a forecast of the cost of the project. This costing also covers the investment plan for the succeeding period (e.g. 1990–1995) which is important for securing technical modernization and for the early preparation of design. It also sets investment limits and serves for the calculation of the rate of return of investments.

Another point among all these economic and administrative problems is the concentration of the hop production in a particular area, as a specialized crop, and the points to be considered for the best organization of work processes according to the production methods to be used.

The economic evaluation of the proposed perspective measures has great importance for all aspects of the economic efficiency of production, i.e. production intensity, labour productivity, investment rate, rentability and expected rate of return of planned investments. The progress of development of production intensity can be expressed by indices based on past and future yield per unit area. The labour productivity can be expressed by a comparison of man hours per ha required in past, with those expected in the future and, similarly, it is possible to compare the investment rate as well as the rentability per ha. In considering investment rate it is effective to use either the total cost or the direct, or specific, cost. To determine the rate of return of investments the same method will be used as is used for the rentability rate. *This can be expressed in percentage terms as the ratio between gross profit and total cost according to the formula:*

$$MR\ (\%) = \frac{HZ}{VN} \cdot 100,$$

where

  MR = rentability rate (%);
  HZ = gross profit (in CKR) (given by subtracting total costs from sales);
  VN = total costs (in CKR).

The conclusion must briefly and clearly evaluate the various parts of the conception and indicate the main problem areas and the measures to be taken for their solution.

SCIENTIFIC AND TECHNICAL PROGRESS IN HOP PRODUCTION

The industrialization of hop production has been achieved through scientific and technical progress. At the start of the socialist regime in Czechoslovakia the total area of hop gardens was 8000 ha, giving an annual harvest of 8 000 000 kg. The organization and management of hop production has undergone great changes with the result that nowadays harvest is about three times the size of that in 1948–1949 though the hopgrowing area has grown by only 25 per cent. The production of hops in Czechoslovakia is of national importance because approximately 70 per cent of the harvest is sent for export and the export requirements for both quantity and quality increase yearly. There is therefore a continuing need for the intensification of hop production.

In recent years yields have considerably increased and are now stabilized at approximately 1.1–1.2 t.ha^{-1} except in years with unfavourable climatical conditions. This, however, is not sufficient for the future, it will be necessary to increase the yield to 1.5–1.65 t.ha^{-1} and, therefore, many scientific and technical problems will need to be solved.

The overall aim and the main direction of scientific research was intended to reach these difficult targets so as to further development of hop production and consumption by means of scientific research. It is presumed that knowledge will be acquired during this prognostic period at such a speed as will ensure that both crop volume and quality will be rapidly increased. This will involve research into new varieties, into crop protectants, into the biochemistry and physiology and nutrition of the hop plant, into agrotechnology and mechanization, and into the economics and administration of hop production. The aim is here to find and to put into practice all such new knowledge so as to further increase the yield and quality of hop production in Czechoslovakia.

New varieties are an important research area for the growth of yield. In comparison with foreign varieties, the Czechoslovak hop has lower yield. Therefore new varieties should provide an increase and in the content of bitter substances, they should have improved resistance to downy mildew and planting material should be virus-free. In order to reach this target a unified classification of morphological properties is required, related to the physio ogical and brewing quality of the breeding material. This will involve a study of Czechoslovak and foreign hop varieties and an investigation of the use of ionizing radiation, chemical mutagens and of polyploidy. At the same time it will be necessary to investigate somatic mutation occurrence in stands of Czechoslovak hops.

272

It cannot be expected that the yields and the bitter substances content of Czechoslovak hops will equal those of other countries, in the short-term, because Czechoslovak regions and Czechoslovak varieties are very different from those elsewhere.

Breeding for downy mildew resistance has made good progress abroad. The disease is important in Czechoslovakia so breeding new varieties is on similar lines to those used in other countries and the results achieved so far show great promise. The same attention and work is also given to virus diseases and to the health of breeding material.

A great deal of attention is paid *to the physiology of yield*, to the importance and contribution of individual physiological processes to crop production, and to the anatomical and morphological structure of various organs in their relationship to external conditions.

An important factor in improving crop yield is the creation of *an efficient systematic protection against pests and diseases* which have a profound effect on the quantity and quality of the product. It will be necessary to pay particular attention to the resistance of pests to insecticides, and to consider the introduction of integrated protection, particularly against sucking insects. The aim will also be to reduce the number of applications of fungicides to control of downy mildew, based on the forecasting of likely outbreaks. This integrated protection and disease forecasting will help to reduce the use of chemical sprays and this will greatly contribute to the protection of the environment in hop-growing regions. It can be presumed that the use of integrated protection will gradually increase the natural resistance of the environment, due to the better survival of the pests predators.

In comparison with the present situation this protective system will give considerable annual savings. If the resistance spectrum of the hop aphid and red spider mite can be defined and compounds selective for the control of these two important hop pests can be found it will soon be possible to introduce integrated protection. However, the application of these rational protection programmes requires sufficient machinery to ensure that the whole planted area is treated within a period of 3–5 days.

Important problems remain to be solved *in the sphere of agrotechnics and the mechanization of hop production*. Their present day level has reduced the required labour to 50 per cent of

Fig. 123. The stamp of the public hop marking establishment at Žatec.

that of the original traditional method. When all divisions of the production process have been mechanized, then the required human work will decrease by a further 25 per cent. The first priority is to complete the mechanization of suspending the wire ropes, to simplify the control of hop growth and to provide the best agrotechnics, nutrition and irrigation.

With respect to *the harvest,* all main parts of the stationary picking machines will be improved and new technology of picking will be investigated. New methods of *drying and post-harvest treatment* will be studied with the aim of preserving the substances important in brewing.

The erection of new hop gardens is associated with the introduction of new large-area structures using pylons of concrete, steel or other materials.

If the present state and the planned parameters of *machinery and technical equipment* produced in Czechoslovakia is compared with the world situation it is clear that the research is in the right direction, producing results of the same, if not of higher, quality. This is demonstrated by the growing export market for our products.

Likewise, the research into *improving nutrition and manuring,* so as to produce increased yields plays an important role in the further intensification of hop production in Czechoslovakia when combined with other agrotechnical advances.

It is most important to determine *the best post-harvest treatment* because this could ensure permanent sales of Czech hops. Qualitative analyses and chemical research need to consider the transformation products of α- and β-bitter acids and their effects on the quality of beer. Of no less importance is the study of spray residues on the quality of treated cones. Both of these research projects need to be speedily concluded.

Permanent attention needs to be given to the complex requirements of the next series of scientific and technical developments. *Economic research* is closely involved in all of the biological and technical problems and is concerned in all of the important economic and administrative problems of hop production in all its aspects and interrelationships.

REFERENCES

Books

AMOS A.: Hop Growing on the Pacific Coast of America. Self-edition. Washington 1913.

ANTIPOVIČ O., VENT L.: Hops (in Czech) in STEHLÍK V. et al.: Breeaing of crops (in Czech). State Agricultural Publishing House, Praha 1959.

BĚLOROSSOVA J. V.: Strain testing of hops (in Russian). Selskochozjajstvenaja akademija im K. A. Timirjazeva, Moscow 1954.

BLATTNÝ C., OSVALD V.: For healthy and first-quality hops (in Czech). Brázda. Praha 1955.

BROOKS N., H ORNER J., LIKENS S.: Hop Production. Agric. Inf. Bull. No 240 USDA, Washington 1961.

BULGAKOV N. J.: The chemistry of brewing process (in Russian). Piščeprom izdatelstvo. Moscow 1954.

BURGESS A. H.: Hops Botany. Cultivation and Utilization. World Crops Books, New York – London 1961.

CAMBIER M.: A propos de l'identification des varietés de houblon. Self-editions. Bruxelles 1958.

ČECH K.: The origin of hop growing and brewery (in Czech). Pivovarské listy (Brewing Industry Journal). Praha 1884 .

DOERLELL E. G.: Die Düngung des Hopfens. Verlag der wissenschaftlichen Anstalten für Bräuindustrie. Prag 1933.

FLEZSYŃSKI T. et al.: Hop growing and processing (in Polish). Rodzial I.P.N.R.

FRIC V.: Technology of continuous harvesting in harvesting center. Thesis for PhD. University of Agriculture. Praha 1965.

FRIC V. et al.: Plant production, State Publishing House of Paedagogical Literature. Praha 1968.

FRUWIRTH C.: Hopfenbau und Hopfenbehandlung. Verlag Paul Parey. Berlin 1928.

FUKSA J., KŘÍŽ J., RYBÁČEK V., SRP A., VENT L.: Hop growing and processing (in Czech) State Agricultural Publishig House. Praha 1956.

GLADISKO S. O., POSESSOR P. F., ŠILO M. P.: Hops (in Ukrainian). Žitomirske oblasne vidavnictvo. Žitomir 1958.

GRASS R.: Hopfen Böhmens. Self-edition. Praha 1904.

GÜNZEL F. V.: Saazer Hopfen. Verlag Franz Viktor Günzel. Žatec 1904.

HAVEL J.: Rationalizing the operations of hop drying based on an ergonomic study. Thesis for PhD. University of Agriculture. Praha 1965.

HLAVÁČEK J., LHOTSKÝ A.: Technologies of brewing industry (in Czech). State Publishing House of Technological Literature. Praha 1972.

HRUŠKA L. et al.: Seeding material and process (in Czech). State Agricultural Publishing House. Praha 1958.

CHODOUNSKÝ F.: Evaluation of hops according to external traits (in Czech). Self-edition. Praha 1898.

KIŠGECI J. et al.: Hop produktion, Poljoprivredni fakultet, Novi Sad, 1984.

KLAPAL M.: Eighty years of hop growing in the town of Tršice (in Czech). Self-edition. Tršice 1941.

KLEČKA A., RYBÁČEK V., VRBENSKÝ V.: Improvement of hop yield (in Czech). Brázda. Praha 1950.

KLITSCH C., RIETZEL P., DIESSEL G.: Die Technik des Hopfenbaues. Neumann Verlag. Radebeul 1956.

KOHLMANN H., KASTNER A.: Der Hopfen. Hopfen Verlag. Wohnzach 1975.

KORFF G., ZATTLER F.: Die Peronosporakrankheit des Hopfens. Landsanstalt für Pflanzenbau und Pflananzenschutz. München 1927.

KŘÍŽ J.: A bionomic and ecologic study of hop aphid (*Phorodon humuli* Schrk.) and elaboration of a protection system (in Czech). Thesis for PhD. University of Agriculture. Praha 1960.

KŘÍŽ J.: New information from the research of viroses applied to crop production (in Czech). Information Center for Science and Technology. Prague 1967.

KUNZ E., SKLÁDAL V.: Hops (in Czech). State Publishing House of Paedagogical Literature. Praha 1956.

KUTINA J.: Growth controllers and their application to agriculture and gardenery. State Agricultural Publishing House. Praha 1988, 2nd edition.

LIBICH V.: Economy and organization of specialized production units (in Czech). Thesis for PhD Research Institute of Economics. Praha 1965.

LINKE W., REBL A.: Der Hopfenbau. Verlag Hanz Carl. Nürnberg 1950.

MATOUŠEK A.: Hop growing and processing (in Czech). Self-edition. Praha 1921.

MOHL A.: Hop growing and processing, Part I, History (in Czech). Neubert's agrarian booksellers. Praha 1906.

MOHL A.: Hop growing and processing, Part II, (in Czech). Neubert's agrarian booksellers. Praha 1924.

NEČIPORČUK J. D.: Agrobiological bases of hop cultivation (in Russian). Izdatelstvo Lvovskogo gosuderstvenogo universiteta. Lvov 1955.

OBENBERGER J., TROJAN V.: Manual of chemical protection of plants (in Czech). State Agricultural Publishing House. Praha 1976.

OSVALD A.: An analytical study of hops from Žatec (in Czech). Ministry of Agriculture of the Czechoslovak Republic. Praha 1929.

OSVALD A.: Studies on hop genetics (in Czech). Union of Hop Growers. Louny 1944.

OSVALD A.: Cultivation of hop (in Czech). Brázda. Praha 1946.

POKORNÝ M., SACHL J., KOSTELNÍK J.: Cultivation of hop with broad spacing. Ministry of Agriculture of the Czechoslovak Republic. Praha 1960.

POPGEORGIJEV G. K.: Hop (in Bulgarian). Zemizdat. Sofia 1962.

RYBÁČEK V.: Hop (in Czech) in Zapletal et al.: Practical agrotechnics (in Czech). State Agricultural Publishing House. Praha 1959.

RYBÁČEK V.: Hops (in Czech) in: Šimon et al.: Plant production (in Czech). State Agricultural Publishing House. Praha 1964.

RYBÁČEK V.: Plant production, Part III (in Czech). State Agricultural Publishing House. 1st edition Praha 1965. 2nd edition Praha 1970.

RYBÁČEK V.: Plant production, Part II (in Czech). State Agricultural Publishing House. Praha 1964 .

RYBÁČEK V.: Some problems of hop biology and their effects on agrotechnics (in Czech). Thesis for PhD. University of Agriculture. Praha 1967.

SALMON F. S., WARE W. M.: The Downy Mildew of the Hop. London 1927.

SAVOV CH., KOVAČEV J.: Hop growing and processing (in Bulgarian). Zemizdat. Sofia 1953.

SRP A.: Effects of organic and mineral manuring on hop yield and quality (in Czech). Thesis for PhD. University of Agriculture. Praha 1966.

ŠNOBL J.: The effects of planting material on young hop plant (in Czech). Thesis for PhD. University of Agriculture. Praha 1975.

ŠTĚPÁNEK P.: Cultivation of hop (in Czech). Self-edition. Milotice 1928.

TOMEŠ J.: Hops (in Czech). A. Reinwart's Publishing House. Praha 1891.

VENT L. et al.: Hop growing and processing (in Czech). State Publishing House of Agricultural Literature. Praha 1963.

WAGNER T.: Autochtonous Hop in Yugoslavia and is Usability for Breeding New Varieties in Comparison with the Hop Varieties grown at Present. Final Report, Podoktorska naloga. Žalec 1974.

ZATTLER F.: Düngung des Hopfens, in Handbuch der Pflanzenernährung und Düngung. Band III. Wien 1965.

ZÁZVORKA V., ZIMA t.: Hop growing and processing (in Czech). State Agricultural Publishing House, Praha 1956.

ZIMA F., ZÁZVORKA V.: Hop growing and processing (in Czech). Ministry of Agriculture of Czechoslovak Republic. Praha 1938.

Scientific journals

Annales Universitatis Mariae Curie-Skłodowska, Lublin-Polonia, 1950.
Annual Report Department of Hop Research, Wye College, 1956–1975.
Annual Report East Malling Research Station, 1956–1962.
Bilten za hmelj a sirak. Novi Sad, 1969–1972.
Bulletin Anc. et Brasserie de l' Université de Louvain, 1934, 1935, 1936, 1937.
EBC-Proceedings, 1963, 1965, 1975.
Hop overzicht van het onderzoek in 1971 en 1972, Beitem-Rumbeke.
Izvestija Timirjazevskoj selskochozjajstvennoj akademii, Moskva, 1977.
Jugoslovenski simpozijum za hmeljarstvo, 1962, 1967, 1973.
Naukovi praci. Ukrajinska doslídna stanica chmeljarstva, Kijiv, 1938, 1961.
Pivovarský sborník, Praha, 1933.
Proceedings of American Society of Brewing Chemists, 1975.
Rostlinná výroba – Chmel. Vědecký časopis ÚVTI, 1962–1978.
Sborník Československé akademie zemědělské, Praha, 1950.
Sborník Vysoké školy zemědělské v Praze, Fàkulta agronomická, 1962–1977.
Trudy. Respublikanskaja naučo-issledovatělskaja chmelovodčeskaja stancija, Moskva, 1954

Scientific letters

Brauwelt, 1925, 1963.
Český chmelař, 1927–1954.
Hmeljar, 1962–1978.
Hopfenbau, 1966–1978.
Hopfen Rundschau, 1954–1978.
Chmelařství, 1955–1978.
Kvasný průmysl, 1960–1978.

SUBJECT INDEX

AUTHOR INDEX

Printed and bound by CPI Group (UK) Ltd, Croydon, CR0 4YY

03/10/2024

01040328-0011